Mastering Essential Math Skills

Pre-Algebra Concepts

Richard W. Fisher

Pre-Algebra Concepts - Item #409
ISBN 10: 0-9666211-9-0 • ISBN 13: 978-0-9666211-9-8

What sets this book apart from other books is its approach. It is not just a math book, but a system of teaching math. Each daily lesson contains three key parts: **Review Exercises**, **Helpful Hints**, and **Problem Solving**. Teachers have flexibility in introducing new topics, but the book provides them with the necessary structure and guidance. The teacher can rest assured that essential math skills in this book are being systematically learned.

This easy-to-follow program requires only fifteen or twenty minutes of instruction per day. Each lesson is concise and self-contained. The daily exercises help students to not only master math skills, but also maintain and reinforce those skills through consistent review - something that is missing in most math programs. Skills learned in this book apply to all areas of the curriculum, and consistent review is built into each daily lesson. Teachers and parents will also be pleased to note that the lessons are quite easy to correct.

This book is based on a system of teaching that was developed by a math instructor over a thirty-year period. This system has produced dramatic results for students. The program quickly motivates students and creates confidence and excitement that leads naturally to success.

Please read the following "How to Use This Book" section and let this program help you to produce dramatic results with your math students.

How to Use This Book

This book is best used on a daily basis. The first lesson should be carefully gone over with students to introduce them to the program and familiarize them with the format. It is hoped that the program will help your students to develop an enthusiasm and passion for math that will stay with them throughout their education.

As you go through these lessons every day, you will soon begin to see growth in the student's confidence, enthusiasm, and skill level. The students will maintain their mastery through the daily review.

Step 1

The students are to complete the review exercises, showing all their work. After completing the problems, it is important for the teacher or parent to go over this section with the students to ensure understanding.

Step 2

Next comes the new material. Use the "Helpful Hints" section to help introduce the new material. Be sure to point out that it is often helpful to come back to this section as the students work independently. This section often has examples that are very helpful to the students.

Step 3

It is highly important for the teacher to work through the two sample problems with the students before they begin to work independently. Working these problems together will ensure that the students understand the topic, and prevent a lot of unnecessary frustration. The two sample problems will get the students off to a good start and will instill confidence as the students begin to work independently.

Step 4

Each lesson has problem solving as the last section of the page. It is recommended that the teacher go through this section, discussing key words and phrases, and also key strategies. Problem solving is neglected in many math programs, and just a little work each day can produce dramatic results.

Step 5

Solutions are located in the back of the book. Teachers may correct the exercises if they wish, or have the students correct the work themselves.

Table of Contents

Sets 4

Integers 6

Positive and Negative Fractions 12

Positive and Negative Decimals 13

Exponents 14

Square Roots 16

Order of Operations 19

Properties of Numbers 21

Scientific Notation 23

Ratios and Proportions 25

Percents 30

Number Theory 37

Number Lines 42

Coordinate Systems 44

Slope 46

Graphing Linear Equations 48

Equations 50

Algebra Word Problems 56

Probability 60

Statistics 63

Final Review of All Pre-Algebra Concepts 67

Solutions 73

Glossary 83

Important Symbols and Tables 89

Review Exercises

Note to students and teachers: This section will include necessary review problems from all types covered in this book. Here are some sample problems with which to get started.

1. $364 + 79 + 716 =$ 2. $705 - 269 =$ 3. $7 \times 326 =$

Helpful Hints

A set is a well-defined collection of objects. A = {1,2,3,4,5} is read, "A is the set whose members are 1, 2, 3, 4, and 5. Each object in a set is called an element or member. **Infinite sets** are sets whose number of members is uncountable. **Example: A = {1,2,3...}** Finite sets are sets whose number of members is countable. **Example: B = {3,4,5}** **Disjoint sets** have no members in common. The **null set** or **empty set** is the set with no members, and is written as { } or Ø. **Equivalent sets** can be paired in a one-to-one correspondence. **Example:** A = {1,2,3,4}

B = {2,3,4,5}

U = the universal set; the set that contains all the members.

$\in$ means "is a member of." $\notin$ means "is not a member of."

Use the information and examples given in the Helpful Hints to answer the following questions. Explain each answer in the space below.

S1. Is A = {2,4,6,8,10} an infinite set?

S2. Is B = {2,4,6...} a finite set?

1. Are A = {1,2,3} and B = {3,4,5} disjoint sets?

2. Are C = {0,1,2,3} and D = {2,4,6,8} equivalent sets?

3. List two disjoint sets.

4. List two equivalent sets.

For 5-10, list the members of each set.

5. {the odd numbers between 2 and 12}

6. {the even numbers less than 13}

7. {the whole numbers between 2 and 10}

8. {the multiples of five between 9 and 32}

9. The members common to A = {1,2,3,4,5} and B = {1,3,5,7}

10. {the whole numbers greater than 7 and less than 13}

1.	
2.	
3.	
4.	
5.	
6.	
7.	
8.	
9.	
10.	
Score	

Problem Solving

In a class of 38 students, one-half are girls. How many girls are there in the class?

4

Review Exercises

1. List two disjoint sets.   2. List two equivalent sets.   3. List an infinite set.

4. List a finite set.   5. Are A = {1,5,10} and B = {5,10,15} disjoint sets? Why?

6. Are C = {2,4,5} and D = {0,1,2,3} equivalent sets? Why?

Helpful Hints

Use the sets below for the following examples that pertain to subset, intersection, and union.

$$A = \{1,2,3\} \qquad B = \{0,1,2,3,4\} \qquad C = \{2,4,6,8,10\}$$

If **A** and **B** are sets and all the members of **A** are members of **B**, then **A** is a **subset** of **B** and is written **A ⊂ B**. **Example:** Is **A ⊂ B**? Yes, because all the members of **A** are members of **B**.

If **A** and **B** are sets then **A intersection B** is the set whose members are included in both sets **A** and **B**, and is written **A ∩ B**. **Example:** Find **A ∩ C**
    A ∩ C = {2} (Two is the only member included in both **A** and **C**.)

If **A** and **B** are sets then **A union B** is the set whose members are included in **A** or **B**, or both **A** and **B**, and is written **A ∪ B**. **Example:** Find **B ∪ C**
    B ∪ C = {0,1,2,3,4,6,8,10} (**B ∪ C** contains all members in **B**, **C**, or both **B** and **C**.)

Use the sets below to answer the questions on this page.
Explain in the space if necessary.

$$A = \{5,6,7\} \qquad B = \{1,2,3,4,5,6,7\} \qquad C = \{1,2,4,5,7,8\} \qquad D = \{1,2,4,6,8,10\}$$

S1. Is A ⊂ B? Why?   S2. Find A ∩ B.

1. Find B ∪ C.   2. Is A ⊂ D? Why?

3. List all subsets of A.   4. Find B ∩ C.
   (Hint: there are seven of them.)

5. Find C ∩ D.   6. Find A ∪ B.

7. Find B ∪ D.   8. Find B ∩ D.

9. Are C and D equivalent sets? Why?

10. Are A and D disjoint sets? Why?

1.
2.
3.
4.
5.
6.
7.
8.
9.
10.
Score

Problem Solving

Three weeks ago you Jose sold seven of his baseball cards from his collection, and last week he bought 12 new cards. If he now has 85 cards, how many did he start with three weeks ago?

Review Exercises

Use A = {1,2,3,5,6},   B = {2,4,8},   and C = {1,2,3,6}   to answer the following questions.

1. Find A ∩ B.

2. Find B ∪ C.

3. Find A ∩ C.

4. Find B ∩ C.

5. Are A and C equivalent sets? Why?

6. Is A an infinite set? Why?

Helpful Hints	**Integers are the set of whole numbers and their opposites.**  Integers to the left of zero are negative and less than zero. Integers to the right of zero are positive and greater than zero. When two integers are on a number line, the one farthest to the right is greater. Hint: When adding integers, always find the sign of the answer first.

Examples: The sum of two negatives is a negative.

$-7 + -5 = -$

(the sign is negative)

$$\begin{array}{r} 7 \\ +5 \\ \hline -12 \end{array}$$

When adding a negative and a positive, the sign is the same as the integer farthest from zero. Then subtract.

$-7 + 9 = +$

(the sign is positive)

$$\begin{array}{r} 9 \\ -7 \\ \hline +2 \end{array}$$

S1. -9 + 12 =

S2. -15 + -6 =

1. -15 + 29 =

2. -12 + -6 =

3. 42 + -56 =

4. -15 + -16 =

5. 8 + 32 =

6. -39 + 76 =

7. -96 + -72 =

8. 73 + -86 =

9. -15 + -19 =

10. 71 + -81 =

1.

2.

3.

4.

5.

6.

7.

8.

9.

10.

Problem Solving

At 3:00 a.m. the temperature was -8°. By 6:00 a.m. the temperature was another -12° colder. What was the temperature at 6:00 a.m.?

Score

Review Exercises

1. -16 + 9 =

2. -6 + 19 =

3. -26 + -13 =

4. -26 + 26 =

5. Carefully define "set."

6. Carefully define "finite set."

Helpful Hints	When adding more than two integers, group the negatives and positives separately, then add.	Examples: -6 + 4 + -5 = -11 + 4 = - (sign is negative)	11 -4 7 = (-7)	7 + -3 + 8 + 6 = -11 + 13 = + (sign is positive)	13 -11 2 = (+2)

S1. -3 + 5 + -6 =

S2. -7 + 6 + -9 + 3 =

1. -3 + -4 + 5 =

2. 7 + -6 + -8 =

3. -15 + 19 + -12 =

4. -6 + 9 + 7 + 4 =

5. -16 + 32 + -18 =

6. -13 + 16 + -8 + 15 =

7. -9 + -7 + -6 =

8. -3 + 7 + -8 + -9 =

9. -32 + 16 + -17 + 8 =

10. -76 + 25 + -33 =

1.	
2.	
3.	
4.	
5.	
6.	
7.	
8.	
9.	
10.	
Score	

Problem Solving

Alice started the week with no money. On Monday she earned $45.00. On Tuesday she spent $27.00. On Wednesday she earned $63.00. On Thursday she spent $26.00. How much money does she have left?

Review Exercises

Use the sets to answer problems 1 through 6.

A = {1,4,8,9,12},   B = {0,5,10,15}, and   C = {9,10,11,15} to answer the following questions.

1. Find A ∩ B.

2. Find A ∪ B.

3. Find B ∩ C.

4. Find A ∪ C.

5. Find A ∪ Ø.

6. Find B ∩ Ø.

| **Helpful Hints** | To subtract an integer means to add to its opposite. | Examples:
-3 - -8 =
-3 + 8 = +
(sign is positive) | 8
- 3
5 =(+5) | 8 - 10 =
8 + -10 = -
(sign is negative) | 10
- 8
2 =(-2) | 6 - -7 =
6 + 7 = +
(sign is positive) | 7
+ 6
13 =(+13) |

S1. -6 - 8 =

S2. 6 - 9 =

1. 3 - -9 =

2. 15 - 18 =

3. -16 - -25 =

4. -16 - 12 =

5. 32 - -14 =

6. -35 - 14 =

7. -6 - 4 =

8. -64 - -53 =

9. -49 - 54 =

10. -63 - -78 =

1.
2.
3.
4.
5.
6.
7.
8.
9.
10.
Score

Problem Solving A boy jumped off a diving board that was 15 feet high. He touched the bottom of the pool that was 12 feet below the surface of the water. How far is it from the diving board to the bottom of the pool?

Review Exercises

1. -72 + 16 =

2. 55 + -33 =

3. -16 + -19 =

4. 7 - 16 =

5. -5 - 6 =

6. -5 - -9 =

Helpful Hints

The product of two integers with different signs is negative. The product of two integers with the same sign is positive. (• means multiply.)

Examples:

$7 • -16 = -$ 16
(sign is negative) $\times 7$
               $112 = \boxed{-112}$

$-8 • -7 = +$ 8
(sign is positive) $\times 7$
            $56 = \boxed{+56}$

When multiplying more than two integers, group them into pairs to simplify. An integer next to parenthesis means to multiply.

Examples:

$2 • -3 \,(-6) =$ 6
$(2 • -3)\,(-6) =$ $\times 6$
$-6\,(-6) = +$ $36 = \boxed{+36}$
(sign is positive)

$-2 • -3 • 4 • -2 =$ 6
$(-2 • -3)\,(4 • -2) =$ $\times 8$
$6 • -8 = -$ $48 = \boxed{-48}$
(sign is negative)

S1. $-3 \times 16 =$

S2. $-18 • 7 =$

1. $-4 • -17 =$

2. $16 \times -4 =$

3. $-24 • -12 =$

4. 23×-16

5. $-23 • 32 =$

6. $(-2)\,(-3)\,(-4) =$

7. $-8\,(-1) • 1\,(-4) =$

8. $4\,(-3) • 2\,(-3) =$

9. $(-3)\,(-2)\,(3)\,(4) =$

10. $10\,(-11)\,(-3) =$

1.	
2.	
3.	
4.	
5.	
6.	
7.	
8.	
9.	
10.	
Score	

Problem Solving

An elevator started on the 28th floor. It went up seven floors, down 13 floors, and up nine floors. On what floor is the elevator located now?

Review Exercises

1. -27 + 16 =

2. -37 + -19 =

3. 7 - 9 =

4. -6 - -8 =

5. 5 • -7 =

6. -2 • -6 • 3 =

Helpful Hints

The quotient of two integers with different signs is negative. The quotient of two integers with the same signs is positive. (HINT: Determine the sign, then divide.)

Examples:

$36 \div -4 = -$ (sign is negative)

$$4\overline{)36}$$
$$\frac{-36}{0} = \boxed{-9}$$

$\dfrac{-123}{-2} = +$ (sign is positive)

$$3\overline{)123}$$
$$\frac{-12\downarrow}{3} = \boxed{+41}$$

Use what you have learned to solve problems like these.

Examples:

$\dfrac{-36 \div -9}{4 \div -2} = \dfrac{4}{-2} = \boxed{-2}$

(sign is negative)

$\dfrac{4 \times -8}{-8 \div 2} = \dfrac{-32}{-4} = \boxed{+8}$

(sign is positive)

S1. -36 ÷ 9 =

S2. $\dfrac{-90}{-15} =$

1. -64 ÷ 4 =

2. -336 ÷ -7 =

3. $\dfrac{-75}{-5} =$

4. 104 ÷ -4 =

5. $\dfrac{54 \div -9}{-18 \div -9} =$

6. $\dfrac{16 \div -2}{-1 \times -4} =$

7. $\dfrac{-75 \div -25}{-3 \div -1} =$

8. $\dfrac{42 \div -2}{-3 \bullet -7} =$

9. $\dfrac{45 \div -5}{-9 \div 3} =$

10. $\dfrac{-56 \div -7}{-36 \div -9} =$

1.
2.
3.
4.
5.
6.
7.
8.
9.
10.
Score

Problem Solving

At midnight the temperature was 7°. By 2:00 a.m. the temperature had dropped 12°. By 4:00 a.m. it had dropped another 6°. What was the temperature at 4:00 a.m.?

Reviewing All Integer Operations

1. -9 + 7 =

2. 9 + -7 =

3. -9 + -7 =

4. -7 + -8 + 14 =

5. -32 + 16 + 21 + -24 =

6. 7 - 9 =

7. 4 - -9 =

8. -3 - 9 =

9. -13 - 14 =

10. 16 - 17 =

11. 3 • -16 =

12. -4 • -19 =

13. 2 (-7) (-4) =

14. -2 • 3 (-4) • 2 =

15. -36 ÷ 4 =

16. -126 ÷ -3 =

17. $\dfrac{-128}{-8} =$

18. $\dfrac{-36 \div 2}{24 \div -4} =$

19. $\dfrac{6 \cdot -3}{-54 \div -6} =$

20. $\dfrac{20 \cdot -3}{-30 \div -10} =$

1.	
2.	
3.	
4.	
5.	
6.	
7.	
8.	
9.	
10.	
11.	
12.	
13.	
14.	
15.	
16.	
17.	
18.	
19.	
20.	

Copyright © 2008, Richard W. Fisher

Review Exercises

Use the following sets to find the answers.

$$A = \{1,3,4,5,9\}, \quad B = \{1,2,4,6\}, \quad \text{and} \quad C = \{1,3,6,7\}$$

1. Find $A \cap B$.

2. Find $B \cup C$.

3. Find $B \cap C$.

4. Find $A \cup B$.

5. Find $A \cup C$.

6. Find $A \cup \varnothing$.

Helpful Hints

The rules for integers apply to positive and negative fractions.

Examples:

$-\dfrac{1}{2} + \dfrac{3}{5} =$

$-\dfrac{5}{10} + \dfrac{6}{10} = +$ (the sign is positive)

$\dfrac{6}{10} - \dfrac{5}{10} = \boxed{\dfrac{1}{10}}$

$-\dfrac{3}{5} + \dfrac{1}{3} =$

$-\dfrac{9}{15} + -\dfrac{5}{15} = -$ (the sign is negative)

$\dfrac{9}{15} + \dfrac{5}{15} = \boxed{-\dfrac{14}{15}}$

$-\dfrac{3}{5} \times 1\dfrac{1}{3} = -$

A negative multiplied or divided by a positive is negative.

$\dfrac{3}{5} \times -\dfrac{3}{2} = \boxed{-\dfrac{9}{10}}$

$-\dfrac{2}{3} \div -\dfrac{1}{2}$

A negative divided by a negative is a positive.

$-\dfrac{2}{3} \times -\dfrac{2}{1} = \dfrac{4}{3} = \boxed{1\dfrac{1}{3}}$

S1. $-\dfrac{1}{5} + \dfrac{1}{2} =$

S2. $\dfrac{1}{2} + -\dfrac{2}{5} =$

1. $\dfrac{1}{2} - \dfrac{3}{4} =$

2. $-\dfrac{2}{3} + -\dfrac{1}{2} =$

3. $-\dfrac{4}{5} \times 2\dfrac{1}{2} = -$

4. $\dfrac{5}{8} + -\dfrac{1}{4} =$

5. $-\dfrac{1}{3} - \dfrac{1}{4} =$

6. $-1\dfrac{2}{3} \times -1\dfrac{1}{2} =$

7. $\dfrac{3}{4} + \dfrac{1}{3} =$

8. $\dfrac{5}{8} + -\dfrac{1}{2} =$

9. $2\dfrac{1}{2} \div -\dfrac{1}{4} =$

10. $-\dfrac{1}{5} + \dfrac{2}{3} =$

1. _____

2. _____

3. _____

4. _____

5. _____

6. _____

7. _____

8. _____

9. _____

10. _____

Problem Solving

There are two sixth-grade classes. One has 35 students and another has 32 students. If a total of 17 sixth graders received A's, how many did not receive A's?

Score

Review Exercises

1. -75 + 16 =   2. -19 - 17 =   3. 16 × -4 =

4. -9 - 19 =   5. -36 ÷ -9 =   6. 2 • -7 × -2 =

Helpful Hints

The rules for integers apply to positive and negative decimals.

Example:

- .71 + .9 = + (the sign is positive)

$$\begin{array}{r} .90 \\ - .71 \\ \hline .19 \end{array}$$

Example:

-2.9 - 3.2 =
-2.9 + - 3.2 = - (the sign is negative)

$$\begin{array}{r} 2.9 \\ + 3.2 \\ \hline -6.1 \end{array}$$

Example:

-.5 × 1.23 = - (a negative multiplied or divided by a positive is a negative)

$$\begin{array}{r} 1.23 \\ × .5 \\ \hline - .615 \end{array}$$

Example:

-3.12 ÷ - .3 = + (a negative multiplied or divided by a negative is a positive)

$$\begin{array}{r} 10.4 \\ .3\overline{)3.12} \end{array}$$

Work the following problems. If necessary, review the rules for integers.

S1. -3.21 + 2.3 =   S2. 5.15 ÷ -.5 =   1. -5.2 - 7.61 =

2. 5.63 + -2.46 =   3. -.7 × 6.12 =   4. 5.9 - -6.23

5. -7.11 ÷ -3 =   6. -.72 + .9 =   7. -2.13 × -.2 =

8. -6.2 + -.73 =   9. 5.2 + -3.19 =   10. -5.112 ÷ .3 =

1.	
2.	
3.	
4.	
5.	
6.	
7.	
8.	
9.	
10.	
Score	

Problem Solving

Anna weighed 120.5 pounds. She lost 7.3 pounds and then gained back 4.8 pounds. How much does she weigh now?

Review Exercises

1. $-.3 + .7 =$

2. $-2.7 + -3.2 =$

3. $3 \times -2.6 =$

4. $-\dfrac{1}{2} + -\dfrac{1}{3} =$

5. $\dfrac{2}{5} + -\dfrac{1}{2} =$

6. $-\dfrac{1}{2} \times -1\dfrac{1}{5} =$

Helpful Hints

In the expression 5^3, the number 5 is called the **base** and the number 3 is called the **power** or **exponent**. The exponent tells how many times the base is to be multiplied by itself. In the example 5^3, you would multiply 5 three times: $5^3 = 5 \times 5 \times 5 = 125$. Negative numbers can have exponents: $(-2)^3 = (-2) \times (-2) \times (-2) = 4 \times (-2) = -8$. Any number to the power of 1 = the number. Any number to the power of 0 = 1.

Examples:

$3^4 = 3 \times 3 \times 3 \times 3$ $(-5)^4 = (-5) \times (-5) \times (-5) \times (-5)$ $5^1 = 5$

$\quad = 9 \times 9$ $\quad = 25 \times 25$ $6^0 = 1$

$\quad = \boxed{81}$ $\quad = \boxed{625}$

S1. $4^2 =$

S2. $-3^3 =$

1. $6^3 =$

2. $5^0 =$

3. $(-2)^4 =$

4. $2^5 =$

5. $7^1 =$

6. $8^3 =$

7. $(-1)^5 =$

8. $5^5 =$

9. $(-5)^3 =$

10. $(-3)^4 =$

1.	
2.	
3.	
4.	
5.	
6.	
7.	
8.	
9.	
10.	
Score	

Problem Solving

A certain number to the third power is equal to eight. What is the number?

Review Exercises

1. $7^2 =$
2. $9^3 =$
3. $(-6)^2 =$

4. $5 + -6 + 8 + -3 =$
5. $7^0 =$
6. $9^1 =$

Many numbers can be written as exponents. **Examples:**

$5 \times 5 \times 5 \times 5 = 5^4$ $(-2) \times (-2) \times (-2) = (-2)^3$

$7 \times 7 \times 7 \times 7 \times 7 = 7^5$ $(-60) \times (-60) \times (-60) = (-60)^3$

$125 = 5^3$
$36 = 6^2$ or $(-6)^2$
$8 = 2^3$
$25 = 5^2$ or $(-5)^2$

Rewrite each of the following as an exponent.

S1. $12 \times 12 \times 12 =$ S2. $27 =$ 1. $2 \times 2 \times 2 \times 2 \times 2 \times 2 =$

2. $(-9) \times (-9) \times (-9) =$ 3. $16 \times 16 \times 16 \times 16 =$ 4. $49 =$

5. $100 =$ 6. $121 =$ 7. $(-1) \times (-1) \times (-1) \times (-1) =$

8. $32 =$ 9. $16 =$ 10. $9 \times 9 \times 9 \times 9 \times 9 \times 9 =$

1.
2.
3.
4.
5.
6.
7.
8.
9.
10.
Score

Problem Solving

A number to the third power is equal to -27. What is the number?

Review Exercises

1. $-36 \div 4 =$

2. $-9 - -6 =$

3. $-\dfrac{1}{3} + -\dfrac{1}{4} =$

4. $-2.7 + 6.3 =$

5. $-3.12 \div 3 =$

6. $\dfrac{3}{4} \times -\dfrac{1}{2} =$

Helpful Hints

$\sqrt{}$ is the symbol for **square root**.

$\sqrt{36}$ is read "the square root of 36."

The answer is the number that when multiplied by itself equals 36.

$\sqrt{36} = 6$, because $6 \times 6 = 36$.

$\sqrt{49} = 7$, because $7 \times 7 = 49$.

$\sqrt{81} = 9$, because $9 \times 9 = 81$.

Find the square roots of the following numbers.

S1. $\sqrt{25} =$ S2. $\sqrt{144} =$ 1. $\sqrt{16} =$

2. $\sqrt{121} =$ 3. $\sqrt{1} =$ 4. $\sqrt{900} =$

5. $\sqrt{100} =$ 6. $\sqrt{400} =$ 7. $\sqrt{169} =$

8. $\sqrt{9} =$ 9. $\sqrt{256} =$ 10. $\sqrt{1,600} =$

1.	
2.	
3.	
4.	
5.	
6.	
7.	
8.	
9.	
10.	
Score	

Problem Solving

The product of -7 and 5 is added to -6.
Find the number.

Review Exercises

1. $6^2 =$

2. $(-2)^3 =$

3. write $6 \times 6 \times 6 \times 6$ as an exponent

4. $\sqrt{64} =$

5. $\sqrt{169} =$

6. $\sqrt{121} =$

Helpful Hints

Use what you have learned about exponents and square roots to solve the following problems.

Examples:

$$\frac{\sqrt{64}}{2^2} = \frac{8}{4} = 2 \qquad\qquad \frac{4^2}{2^3} = \frac{16}{8} = 2$$

$$\frac{3^3}{\sqrt{9}} = \frac{27}{3} = 9 \qquad\qquad \sqrt{36} \times 3^3 = 6 \times 27 = 162$$

Solve each of the following.

S1. $\sqrt{16} \times 3^2 =$

S2. $\dfrac{4^3}{\sqrt{64}} =$

1. $\dfrac{\sqrt{100}}{\sqrt{25}} =$

2. $\dfrac{\sqrt{64}}{(2^3)} =$

3. $\dfrac{5^3}{\sqrt{25}} =$

4. $2^3 \times \sqrt{121} =$

5. $3^2 \times 4^2 =$

6. $\dfrac{2^3 \times 4^2}{\sqrt{4}} =$

7. $\sqrt{81} \times \sqrt{36} =$

8. $\dfrac{2^4}{\sqrt{16}} =$

9. $2^2 \times 3^2 \times \sqrt{16} =$

10. $\dfrac{3^4}{\sqrt{81}} =$

1.	
2.	
3.	
4.	
5.	
6.	
7.	
8.	
9.	
10.	
Score	

Problem Solving

5 to the second power added to the square root of what number is equal to 34?

Reviewing Exponents and Square Roots

For 1-6, rewrite each as an exponent.

1. $13 \times 13 \times 13 \times 13 =$

2. $2 \times 2 \times 2 \times 2 \times 2 \times 2 \times 2 =$

3. $64 =$

4. $(-2) \times (-2) \times (-2) \times (-2) =$

5. $8 =$

6. $100 =$

For 7-12, find the square root of each number.

7. $\sqrt{16} =$

8. $\sqrt{64} =$

9. $\sqrt{16 + 9} =$

10. $\sqrt{400}$

11. $\sqrt{9}$

12. $\sqrt{4} \times 9 =$

Solve each of the following

13. $\sqrt{36} + 4^2 =$

14. $\dfrac{\sqrt{64}}{\sqrt{4}} =$

15. $6^2 + 7^2 =$

16. $(4^2) \times (5^2) =$

17. $\sqrt{49} \times \sqrt{81} =$

18. $3^2 \times 5^2 \times \sqrt{9} =$

19. $\dfrac{5^3}{5} =$

20. $\dfrac{\sqrt{100} \times \sqrt{25}}{\sqrt{25}} =$

1.
2.
3.
4.
5.
6.
7.
8.
9.
10.
11.
12.
13.
14.
15.
16.
17.
18.
19.
20.

Review Exercises

1. $7^2 =$

2. $\sqrt{36} =$

3. $-9 - -7 =$

4. $16 + -72$

5. $\dfrac{16 \div -2}{-4 \times -2} =$

6. $7^2 - 5^2 =$

Helpful Hints	It is necessary to follow the correct **order of operations** when simplifying an expression. 1. Evaluate within grouping symbols. 2. Eliminate all exponents. 3. Multiply and divide in order from left to right. 4. Add and subtract in order from left to right.

Examples:

$3^2 (3 + 5) + 3$
$= 3^2 (8) + 3$
$= 9 (8) + 3$
$= 72 + 3$
$= 75$

*A number next to a grouping symbol means multiply.

Sometimes there are no grouping symbols.

$4 + 12 \times 3 - 8 \div 4$
$= 4 + 36 - 2$
$= 40 - 2$
$= 38$

$3 (2 + 1) = 3 \times (2 + 1)$

Solve each of the following. Be sure to follow the correct order of operations.

S1. $5 + 9 \times 3 - 4 =$

S2. $8 + 3^2 \times 4 - 6 =$

1. $4 (6 + 2) - 5^2 =$

2. $(14 - 6) + 56 \div 2^3 =$

3. $5^2 + (15 + 3) \div 2 =$

4. $7 \times 4 - 9 \div 3 =$

5. $(3 \times 12) \div (9 \div 3) =$

6. $5^2 + 2^3 - 2 \times 3 =$

7. $12 - 6 \div 3 + 4 =$

8. $3^2 - 2^3 + 6 \div 2 =$

9. $(3 + 8 \div 2) \times (2 \times 6 \div 3) =$

10. $9 + [(4 + 5) \times 3] =$

1.	
2.	
3.	
4.	
5.	
6.	
7.	
8.	
9.	
10.	
Score	

Problem Solving

A running back gained 12 yards. The next play he lost 18 yards, and on the third play he gained five yards. What was his net gain or net loss?

Review Exercises

Use the following sets to find the answers.

$$A = \{1,5,7,8,9\}, \quad B = \{2,4,6,8,10\}, \text{ and } \quad C = \{1,2,4,5\}$$

1. Find $B \cap C$.

2. Find $A \cap B$.

3. Find $C \cap \emptyset$.

4. Find $A \cup C$.

5. Find $B \cup C$.

6. Find $(A \cap B) \cup C$.

Helpful Hints

*Remember the correct order of operations:
1. Evaluate within grouping symbols.
2. Eliminate all exponents.
3. Multiply and divide in order from left to right.
4. Add and subtract in order from left to right.

Examples:

$$14 \div 2 \times 3 + 4^2 - 1$$
$$= 14 \div 2 \times 3 + 16 - 1$$
$$= 7 \times 3 + 16 - 1$$
$$= 21 + 16 - 1$$
$$= 37 - 1$$
$$= 36$$

$$\frac{5(8-3) - 2^2}{3 + 2(3^2 - 7)} = \frac{25 - 4}{3 + 4}$$
$$= \frac{5(5) - 2^2}{3 + 2(9-7)} = \frac{21}{7}$$
$$= \frac{5(5) - 4}{3 + 2(2)} = 3$$

S1. $\{(2+4) \times 3 + 2\} \div 5 =$

S2. $\dfrac{7^2 - (-5+9)}{2(4^2 - 12) - 3} =$

1. $4^3 - 7(2+3) =$

2. $\dfrac{(12-3) + 3^2}{-7 + 2(4+1)} =$

3. $\dfrac{4^2 + 12}{5 + 3(2+1)} =$

4. $6(-3+9) + -4 =$

5. $\dfrac{10 + (2 + -6)}{4(2^3 - 6) + -2} =$

6. $63 \div 7 - 3 \times 2 + 4 =$

7. $3\{(2+7) \div 3 + 7\} \div 5 =$

8. $(12 + -3) + 75 \div 5^2 =$

9. $20 - 3^2 - 5 \times 2 + 6 =$

10. $3^2 + 2^3 + 14 \div 2 =$

1.

2.

3.

4.

5.

6.

7.

8.

9.

10.

Problem Solving

John started the week with $64. Each day, Monday through Friday, he spent $7 for lunch. How much money did he have left at the end of the week?

Score

20

Review Exercises

1. $3^3 =$

2. $5 + 3 \times 7 + 2 =$

3. $-7 - 6 =$

4. Carefully define "set."

5. $\frac{1}{2} \times -2\frac{1}{2} =$

6. $-.91 + .5 =$

Helpful Hints	For any real numbers a, b, and c, the following properties are true:		**Examples:**
	1. Identity Property of Addition	$0 + a = a$	$0 + 2 = 2$
	2. Identity Property of Multiplication	$1 \times a = a$	$1 \times 7 = 7$
	3. Inverse Property of Addition	$a + (-a) = 0$	$5 + (-5) = 0$
	4. Inverse Property of Multiplication	$a \times \frac{1}{a} = 1 \quad (a \neq 0)$	$6 \times \frac{1}{6} = 1$
	5. Associative Property of Addition	$(a + b) + c = a + (b + c)$	$(2 + 3) + 4 = 2 + (3 + 4)$
	6. Associative Property of Multiplication	$(a \times b) \times c = a \times (b \times c)$	$(2 \times 3) \times 4 = 2 \times (3 \times 4)$
	7. Commutative Property of Addition	$a + b = b + a$	$5 + 6 = 6 + 5$
	8. Commutative Property of Multiplication	$a \times b = b \times a$	$4 \times 3 = 3 \times 4$
	9. Distributive Property	$a \times (b + c) = a \times b + a \times c$	$5 \times (3 + 2) = 5 \times 3 + 5 \times 2$

Name the property that is illustrated.

S1. $7 + 9 = 9 + 7$ S2. $3 \times (7 + 4) = 3 \times 7 + 3 \times 4$ 1. $7 + (-7) = 0$

2. $3 \times (4 \times 5) = (3 \times 4) \times 5$ 3. $0 + (-6) = -6$ 4. $5 \times \frac{1}{5} = 1$

5. $9 + (6 + 5) = 9 + (5 + 6)$ 6. $9 \times 7 = 7 \times 9$ 7. $(6 + 5) + 7 = 6 + (5 + 7)$

8. $1 \times \frac{7}{8} = 7$ 9. $3 \times 2 + 3 \times 4 = 3 \times (2 + 4)$ 10. $16 + (-16) = 0$

1. _____

2. _____

3. _____

4. _____

5. _____

6. _____

7. _____

8. _____

9. _____

10. _____

Problem Solving

Five times a certain number is equal to 95. Find the number.

Score _____

Review Exercises

1. $3 + 6 \times 2 - 4 =$
2. $3(5 + 2) - 2^2 =$
3. $3 \times 4 - 6 \div 3 =$

4. $2^2 + 3^3 - 2 \times 4 =$
5. $(4 + 4 \div 2) \times (2 \times 10 \div 2) =$
6. $3[(4 + 3) \times 2] =$

Helpful Hints	Use what you have learned to solve the following problems. **Example:** Use the indicated property to complete the statement with the correct answer.
	Inverse Property of Addition: $27 + (\ \ ) = 0$ answer: -27
	Distributive Property: $4(5 + 7) =$ answer: $4 \times 5 + 4 \times 7$

Use the indicated property to complete the statement with the correct answer.

S1. Associative Property of Addition: $(3 + 7) + 9 =$

S2. Commutative Property of Multiplication: $7 \times 15 =$

1. Inverse Property of Multiplication: $9 \times (\ \ ) = 1$

2. Distributive Property: $3 \times (6 + 2) =$

3. Commutative Property of Addition: $9 + 12 =$

4. Associative Property of Multiplication: $3 \times (9 \times 5) =$

5. Distributive Property: $3 \times 5 + 3 \times 7 =$

6. Inverse Property of Addition: $9 + (\ \ ) = 0$

7. Identity Property of Multiplication: $7 \times (\ \ ) = 7$

8. Inverse Property of Multiplication: $\dfrac{1}{5}(\ \ ) = 1$

9. Associative Property of Addition: $3 + (5 + 6) =$

10. Distributive Property: $6 \times (4 + (-2)) =$

1.

2.

3.

4.

5.

6.

7.

8.

9.

10.

Score

Problem Solving	Mr. Andrews rents a car for one day. He pays $30 per day for the rental plus $.30 per mile he drives. How much will the total price of the rental car be if he drives 40 miles?

Review Exercises

1. $\sqrt{100}$

2. $5^3 =$

3. $\dfrac{\sqrt{36} \times \sqrt{49}}{2} =$

4. $-3 \times -5 \times -6 =$

5. $-2\dfrac{1}{2} \div -\dfrac{1}{2} =$

6. $-\dfrac{1}{3} + -\dfrac{1}{2} =$

Helpful Hints

Scientific notation is used to express very large and very small numbers. A number in scientific notation is expressed as the product of two factors. The first factor is a number between 1 and 10 and the second factor is a power of 10 as in the examples 2.346×10^5 and 3.976×10^{-7}.

Example for a large number: Change 157,000,000,000 to scientific notation. Move the decimal between the 1 and the 5. Since the decimal has moved 11 places to the **left**, the answer is 1.57×10^{11}.

Example for a small number: Change .0000000468 to scientific notation. Move the decimal between the 4 and the 6. Since the decimal has moved eight places to the **right**, the answer is 4.68×10^{-8}.

Change the following numbers to scientific notation.

S1. 2,360,000,000

S2. .000000149

1. 653,000,000,000

2. 159,700

3. 106,000,000

4. .000007216

5. 1,096,000,000

6. .001963

7. .00000000016

8. .0000000008

9. 7,000,000,000,000

10. .0000001287

1.
2.
3.
4.
5.
6.
7.
8.
9.
10.
Score

Problem Solving

Light travels at 186,000 miles per second. Write this speed in scientific notation.

Review Exercises

1. Change 123,000 to scientific notation.

2. Change .000321 to scientific notation.

3. Which property of numbers is illustrated?
$3 \times 5 + 3 \times 7 = 3 \times (5 + 7)$

4. $-9 - 8 =$

5. $\dfrac{36 \div -3}{-16 \div -4} =$

6. $2^3 + 3^3 =$

Helpful Hints

It is easy to change numbers in scientific notation to conventional numbers.

Examples:

Change 3.458×10^8 to a conventional number.
Move the decimal eight places to the right. The answer is 345,800,000.

Change 4.5677×10^{-7} to a conventional number.
Move the decimal seven spaces to the left. The answer is .00000045677.

Change each number in scientific notation to a conventional number.

S1. 7.032×10^6

S2. 5.6×10^{-5}

1. 2.3×10^5

2. 9.13×10^{-8}

3. 1.2362×10^{-5}

4. 5.17×10^{11}

5. 1.127×10^3

6. 3.012×10^{-3}

7. 6.67×10^6

8. 2.1×10^4

9. 7×10^{-8}

10. 8×10^6

1.

2.

3.

4.

5.

6.

7.

8.

9.

10.

Score

Problem Solving

The distance to the sun is approximately 9.3×10^7 miles. Change this distance to a conventional number.

24

Review Exercises

1. Change 123,000 to scientific notation.

2. Change .0000056 to scientific notation.

3. Change 2.76×10^6 to a conventional number.

4. Change 3.75×10^{-5} to to a conventional number.

5. List two equivalent sets.

6. List two disjoint sets.

Helpful Hints	A **ratio** compares two numbers or groups of objects. Example: ⃝ ⃝ ⃝ For every three circles there are four squares. ☐ ☐ ☐ ☐ The ratio can be written in the following ways: 3 to 4, 3 : 4, and $\frac{3}{4}$. Each of these is read as "three to four."

*Ratios are often written in fraction form. The first number mentioned is the numerator. Ratios that are expressed as fractions can be reduced to lowest terms.

Write each of the following ratios as a fraction reduced to lowest terms.

S1. 5 nickels to 3 dimes

S2. 18 horses to 4 cows

1. 7 to 2

2. 6 children to 5 adults

3. 30 books to 25 pencils

4. 15 bats to 3 balls

5. 24 to 20

6. 16 to 12

7. 7 dimes to 3 pennies

8. 6 chairs to 4 desks

9. 4 cats to 8 dogs

10. 9 : 3

1.
2.
3.
4.
5.
6.
7.
8.
9.
10.
Score

Problem Solving

A team won 24 games and lost 10. Write the ratio of games won to games lost as a fraction reduced to lowest terms.

Review Exercises

1. Write .00027 in scientific notation.

2. Write 2,916,000 in scientific notation.

3. Write 7.21×10^5 as a conventional number.

4. Write 6.23×10^{-5} as a conventional number.

5. $(-3) \times 2 \times (-5) =$

6. $-.264 \div .2 =$

Helpful Hints

Two equal ratios can be written as a **proportion**.

Example: $\frac{4}{6} = \frac{2}{3}$ In a proportion, the cross products are equal.

Examples: Is $\frac{3}{4} = \frac{5}{6}$ a proportion? To find out, cross multiply.

$3 \times 6 = 18$, $4 \times 5 = 20$, $18 \neq 20$. It is not a proportion.

Is $\frac{6}{9} = \frac{8}{12}$ a proportion? To find out, cross multiply.

$6 \times 12 = 72$, $9 \times 8 = 72$, $72 = 72$. It is a proportion.

Cross multiply to determine whether each of the following is a proportion.

S1. $\frac{2}{5} = \frac{6}{15}$

S2. $\frac{18}{24} = \frac{4}{5}$

1. $\frac{7}{14} = \frac{3}{6}$

2. $\frac{5}{3} = \frac{14}{9}$

3. $\frac{18}{2} = \frac{27}{3}$

4. $\frac{4}{5} = \frac{12}{15}$

5. $\frac{15}{20} = \frac{6}{8}$

6. $\frac{5}{2} = \frac{11}{4}$

7. $\frac{2}{5} = \frac{12}{30}$

8. $\frac{3}{1.3} = \frac{9}{3.5}$

9. $\frac{\frac{1}{4}}{4} = \frac{\frac{1}{2}}{8}$

10. $\frac{5}{8} = \frac{6}{7}$

1.	
2.	
3.	
4.	
5.	
6.	
7.	
8.	
9.	
10.	
Score	

Problem Solving

A whole number to the power of three, added to five, equals 13. Find the whole number.

Review Exercises

1. Write 25 to 15 as a fraction in lowest terms.

2. Is $\dfrac{4}{5} = \dfrac{8}{10}$ a proportion? Why?

3. Is $\dfrac{2}{5} = \dfrac{5}{7}$ a proportion? Why?

4. $\sqrt{49} + 3^2 =$

5. $4^3 - 2^4 =$

6. $-225 + 500 =$

Helpful Hints

It is easy to find the missing number in a proportion.

Examples: Solve each proportion.

$\dfrac{4}{n} = \dfrac{2}{3}$ First, cross multiply: $2 \times n = 4 \times 3$
$2 \times n = 12$

Next, divide 12 by 2: $\dfrac{6}{2\,\overline{)12}}$ $\boxed{n = 6}$

$\dfrac{4}{5} = \dfrac{y}{7}$

First, cross multiply: $5 \times y = 4 \times 7$
$5 \times n = 28$

Next, divide 28 by 5: $\dfrac{5\,\frac{3}{5}}{5\,\overline{)28}}$ $\boxed{y = 5\frac{3}{5}}$

Find the missing number in each proportion.

S1. $\dfrac{3}{15} = \dfrac{n}{5}$

S2. $\dfrac{4}{7} = \dfrac{x}{28}$

1. $\dfrac{n}{4} = \dfrac{12}{16}$

2. $\dfrac{x}{40} = \dfrac{5}{100}$

3. $\dfrac{1}{3} = \dfrac{14}{y}$

4. $\dfrac{n}{4} = \dfrac{8}{5}$

5. $\dfrac{15}{20} = \dfrac{n}{8}$

6. $\dfrac{7}{n} = \dfrac{3}{9}$

7. $\dfrac{27}{3} = \dfrac{n}{2}$

8. $\dfrac{n}{2} = \dfrac{7}{5}$

9. $\dfrac{n}{1.4} = \dfrac{6}{7}$

10. $\dfrac{7}{n} = \dfrac{21}{7}$

1.

2.

3.

4.

5.

6.

7.

8.

9.

10.

Problem Solving

The temperature at midnight is -12°. By 6:00 a.m., the temperature has dropped another 20°. What is the temperature at 6:00 a.m.?

Score

Review Exercises

1. Is $\frac{3}{4} = \frac{9}{12}$ a proportion? Why?

2. Solve the proportion:

   $\frac{n}{12} = \frac{5}{2}$

3. Solve the proportion:

   $\frac{5}{6} = \frac{10}{n}$

4. Write 234,000,000 in scientific notation.

5. Write .00235 in scientific notation.

6. Write 7.2×10^5 as as a conventional number.

Helpful Hints

Ratios and **proportions** can be used to solve problems.

Example: A car can travel 384 miles in six hours. How far can the car travel in eight hours?

First set up a proportion. $\frac{384 \text{ miles}}{6 \text{ hours}} = \frac{n \text{ miles}}{8 \text{ hours}}$ Next, divide by six.

Next, cross multiply. $6 \times n = 8 \times 384$

$6 \times n = 3{,}072$

$\begin{array}{r} 512 \\ 6\,\overline{)3072} \end{array}$ $n = 512$

The car can travel 512 miles in eight hours.

Use a proportion to solve each problem.

S1. A car can travel 85 miles on five gallons of gas. How far can the car travel on 12 gallons of gas?

S2. If two pounds of beef cost $4.80, how much will five pounds cost?

1. A car can travel 100 miles on five gallons of gas. How many gallons will be needed to travel 40 miles?

2. Two pounds of chicken cost $7. How much will five pounds cost?

3. In a class, the ratio of boys to girls is four to three. If there are 20 boys in the class, how many girls are there?

4. A runner takes three hours to go 24 miles. At this rate, how far could he run in five hours?

5. Seven pounds of nuts cost $5. How many pounds of nuts can you buy with $2?

1.
2.
3.
4.
5.
Score

Problem Solving

At 6:00 a.m. the temperature was -16°. By noon the temperature had risen 28°. What was the temperature at noon?

Reviewing Ratios and Proportions

For 1-3, write each ratio as a fraction reduced to lowest terms.

1. 12 to 4　　　　　　**2.** 24 to 10　　　　　　**3.** 16 to 6

For 4-6, determine whether each is a proportion and why.

4. $\dfrac{12}{15} = \dfrac{24}{30}$　　　　**5.** $\dfrac{7}{8} = \dfrac{8}{9}$　　　　**6.** $\dfrac{5}{3} = \dfrac{15}{9}$

For 7-15, solve each proportion.

7. $\dfrac{12}{15} = \dfrac{n}{5}$　　　　**8.** $\dfrac{1}{3} = \dfrac{11}{n}$　　　　**9.** $\dfrac{1}{20} = \dfrac{n}{100}$

10. $\dfrac{5}{7} = \dfrac{25}{n}$　　　　**11.** $\dfrac{3}{4} = \dfrac{n}{6}$　　　　**12.** $\dfrac{15}{20} = \dfrac{n}{12}$

13. $\dfrac{10}{100} = \dfrac{2}{n}$　　　**14.** $\dfrac{x}{5} = \dfrac{9}{15}$　　　**15.** $\dfrac{3}{16} = \dfrac{n}{48}$

For 16-20, use a proportion to solve each problem.

16. If four pounds of pork cost $4.80, how much will seven pounds cost?

17. In a class the ratio of girls to boys is two to three. If there are 20 girls, how many boys are in the class?

18. A cyclist can travel 42 miles in three hours. How far can he travel in five hours?

19. A car can travel 120 miles on five gallons of gas. How many gallons will be needed to travel 48 miles?

20. If six pounds of nuts cost $18, how many pounds of nuts can you buy with $12?

1.
2.
3.
4.
5.
6.
7.
8.
9.
10.
11.
12.
13.
14.
15.
16.
17.
18.
19.
20.

Review Exercises

1. Solve the proportion:
$$\frac{7}{6} = \frac{n}{18}$$

2. Solve the proportion:
$$\frac{n}{3} = \frac{6}{5}$$

3. $3 \times 2 + 6 \div 2 =$

4. $4^2 + (5 \times 2) \div 5 =$

5. $4^2 + 2^2 + 12 \div 2 =$

6. $5(-2 + -6) + 7 =$

Helpful Hints

Percent means "**per hundred**" or "**hundredths.**"

Percents can be expressed as decimals and as fractions.
The fraction form may sometimes be reduced to its lowest terms.

Examples: $25\% = .25 = \frac{25}{100} = \frac{1}{4}$ $\qquad$ $8\% = .08 = \frac{8}{100} = \frac{2}{25}$

Change each percent to a decimal and to a fraction reduced to its lowest terms.

S1. $20\% = $ _____

S2. $9\% = $ _____

1. $16\% = $ _____

2. $6\% = $ _____

3. $75\% = $ _____

4. $40\% = $ _____

5. $1\% = $ _____

6. $45\% = $ _____

7. $12\% = $ _____

8. $5\% = $ _____

9. $50\% = $ _____

10. $13\% = $ _____

1.	
2.	
3.	
4.	
5.	
6.	
7.	
8.	
9.	
10.	
Score	

Problem Solving

95% of the students enrolled in a school are present.
What fraction are present? (Reduce to lowest terms.)

Review Exercises

1. Change 80% to a decimal.   2. Change 7% to a decimal.   3. Change 25% to a fraction reduced to lowest terms.

4. 156
   × .7

5. 400
   × .32

6. 300
   × .06

Helpful Hints	To find the **percent of a number**, you may use either fractions or decimals. Use what is the most convenient.

Example: Find 25% of 60.

$.25 \times 60$

$$\begin{array}{r} 60 \\ \times\ .25 \\ \hline 300 \\ 120 \\ \hline 15.00 \end{array}$$

OR

$$\frac{25}{100} = \frac{1}{4}$$

$$\frac{1}{4} \times \frac{60^{15}}{1} = \frac{15}{1} = \boxed{15}$$

S1. Find 70% of 25.   S2. Find 50% of 300.   1. Find 6% of 72.

2. Find 60% of 85.   3. Find 25% of 60.   4. Find 45% of 250.

5. Find 10% of 320.   6. Find 40% of 200.   7. Find 4% of 250.

8. Find 90% of 240.   9. Find 75% of 150.   10. Find 2% of 660.

1.

2.

3.

4.

5.

6.

7.

8.

9.

10.

Problem Solving	Arlene took a test with 40 questions. If she got a score of 85% correct, how many problems did she get correct?	Score

Review Exercises

1. Find 15% of 310.

2. Find 20% of 120.

3. $8\overline{)6}$

4. Change .7 to a percent.

5. Find .9 of 150.

6. $.05\overline{)30}$

Helpful Hints

When finding the **percent**, first write a fraction, change the fraction to a decimal, and then change the decimal to a percent.

Examples: 4 is what percent of 16?

$\frac{4}{16} = \frac{1}{4}$

$\begin{array}{r} .25 = \boxed{25\%} \\ 4\overline{)1.00} \\ -\ 8\downarrow \\ \hline 20 \\ -\ 20 \\ \hline 0 \end{array}$

5 is what percent of 25?

$\frac{5}{25} = \frac{1}{5}$

$\begin{array}{r} 20 = \boxed{20\%} \\ 5\overline{)1.00} \\ -\ 1\ 0 \\ \hline 00 \end{array}$

S1. 3 is what percent of 12?

S2. 15 is what percent of 20?

1. 7 is what percent of 28?

2. 20 is what percent of 25?

3. 40 = what percent of 80?

4. 18 is what percent of 20?

5. 12 is what percent of 20?

6. 9 is what percent of 12?

7. 15 = what percent of 20?

8. 24 is what percent of 32?

9. 400 is what percent of 500?

10. 19 is what percent of 20?

1.	
2.	
3.	
4.	
5.	
6.	
7.	
8.	
9.	
10.	
Score	

Problem Solving

A rancher had 800 cows. He sold 600 of them. What percent of the cows did he sell?

Review Exercises

1. Find 4% of 80.

2. Find 40% of 80.

3. Twelve is what percent of 16?

4. 45 is what percent of 50?

5.

$52 - 1.96 =$

6. $.06 \overline{) 12}$

Helpful Hints	To find the **whole** when the **part** and the **percent** are known, simply change the equal sign (" = ") to the division sign (" ÷ ").

Examples: 6 = 25% of what number?   Twelve is 80% of what?
6 ÷ 25% (Change = to ÷.)   12 ÷ 80% (Change = to ÷.)
6 ÷ .25 (Change % to decimal.)   12 ÷ .8 (Change % to decimal.)

$.25 \overline{) 6.00}$ ⟨24.⟩   * Be careful to move decimal points properly. $.8 \overline{) 12.0}$ ⟨15.⟩

S1. 5 = 25% of what?

S2. Six is 20 % of what?

1. 12 = 25% of what?

2. 32 = 40 % of what?

3. Five is 20% of what?

4. 3 = 75% of what?

5. Twelve is 80% of what?

6. 8 = 40% of what?

7. 15 is 25% of what?

8. Fifteen is 20% of what?

9. 9 is 20% of what?

10. 25 is 20% of what?

1.	
2.	
3.	
4.	
5.	
6.	
7.	
8.	
9.	
10.	
Score	

Problem Solving

There are 15 girls in a class. If this is 60% of the class, how many students are there in the class?

Review Exercises

1. Change $\frac{72}{100,000}$ to a decimal.

2. Change 2.0019 to a mixed numeral.

3. Change $\frac{9}{15}$ to a percent.

4. $\frac{3}{5} \times 25 =$

5. $8\overline{).168}$

6. $.03\overline{)2.4}$

Helpful Hints

Use what you have learned to solve the following word problems. **Examples:**

A man earns $300 and spends 40% of it. How much does he spend?

Find 40% of 300.

$$\begin{array}{r} 300 \\ \times\ .4 \\ \hline 120 \end{array}$$

He spends $120.

In a class of 25 students, 15 are girls. What % are girls?

15 = what % of 25

$$\frac{15}{25} = \frac{3}{5}$$

$$5\overline{)3.00}^{.60}$$

60% are girls.

Five students got A's on a test. This is 20% of the class. How many are in the class?

5 = 20% of what?
5 ÷ .2

$$.2\overline{)5.0}^{25.}$$

25 are in the class.

S1. On a test with 25 questions, Al got 80% correct. How many questions did he get correct?

S2. A player took 12 shots and made 9. What percent did he make?

1. A girl spent $5. This was 20% of her earnings. How much were her earnings?

2. Buying a $8,000 car requires a 20% down payment. How much is the down payment?

3. 3 = 10% of what?

4. A team played 20 games and won 18. What % did they win?

5. A farmer sold 50 cows. If this was 20% of his herd, how many cows were in his herd?

6. 20 = 80% of what?

7. Paul wants a bike that costs $400. If he has saved 60% of this amount, how much has he saved?

8. There are 400 students in a school. Fifty-five percent are girls. How many boys are there?

9. 12 is what % of 60?

10. Kelly earned 300 dollars and put 70% of it into the bank. How much did she put into the bank?

1.
2.
3.
4.
5.
6.
7.
8.
9.
10.
Score

Problem Solving

Nacho's monthly income is $4,800. What is his annual income? (Hint: How many months are there in a year?)

Review Exercises

1. $7.68 + 19.7 + 5.364 =$

2. $\begin{array}{r} 7.123 \\ -\ 4.765 \\ \hline \end{array}$

3. $\begin{array}{r} 3.14 \\ \times\ 7 \\ \hline \end{array}$

4. $\begin{array}{r} .208 \\ \times\ .06 \\ \hline \end{array}$

5. $3\overline{)1.44}$

6. $.15\overline{)1.215}$

Helpful Hints

Use what you have learned to solve the following problems.

*Refer to the examples on the previous page if necessary.

S1. Find 20% of 150.

S2. 6 is 20% of what?

1. 8 is what % of 40?

2. Change $\frac{18}{20}$ to a percent.

3. A school has 600 students. If 5% are absent, how many students are absent?

4. A quarterback threw 24 passes and 75% of them were caught. How many were caught?

5. Riley has 250 marbles in his collection. If 50 of them are red, what percent of them are red?

6. A team played 60 games and won 45 of them. What % did they win?

7. There are 50 sixth graders in a school. This is 20% of the school. How many students are in the school?

8. A coat is on sale for $20. This is 80% of the regular price. What is the regular price?

9. Steve has finished $\frac{3}{5}$ of his test. What percent of the test has he finished?

10. Alex wants to buy a computer priced at $640. If sales tax is 8%, what is the total cost of the computer?

1.
2.
3.
4.
5.
6.
7.
8.
9.
10.
Score

Problem Solving

Ann took five tests and scored a total of 485 points. What was her average score?

Reviewing Percents

Change numbers 1 - 5 to percents.

1. $\dfrac{13}{100} =$ 2. $\dfrac{3}{100} =$ 3. $\dfrac{7}{10} =$ 4. $.19 =$ 5. $.6 =$

Change numbers 6 - 8 to a decimal and a fraction expressed in lowest terms.

6. $8\% = .\underline{\quad} = \underline{\quad}$ 7. $18\% = .\underline{\quad} = \underline{\quad}$ 8. $80\% = .\underline{\quad} = \underline{\quad}$

Solve the following problems. Label the word problem answers.

9. Find 3% of 74.

10. Find 40% of 320.

11. 20 is what percent of 25?

12. 15 is what percent of 20?

13. $3 = 25\%$ of what?

14. $15 = 20\%$ of what?

15. Change $\dfrac{16}{20}$ to a %.

16. Change $\dfrac{3}{5}$ to a percent.

17. 640 students attend Lincoln School. If 40% of the students are girls, how many girls attend Lincoln School?

18. A team played 40 games. If they won 65% of them, how many games did the team win?

19. A pitcher threw 40 pitches. If 30 were strikes, what percent were strikes?

20. Thirty students attended an assembly. This was 20% of the seventh grade. How many students are there in the seventh grade?

1.
2.
3.
4.
5.
6.
7.
8.
9.
10.
11.
12.
13.
14.
15.
16.
17.
18.
19.
20.

Review Exercises

1. Change .7 to a percent.

2. Change $\frac{4}{5}$ to a percent.

3. Change .12 to a fraction reduced to lowest terms.

4. Find 6% of 200.

5. Three is what percent of 12?

6. 5 = 20% of what?

Helpful Hints

A **factor** of a whole number is a whole number that divides into it evenly, without a remainder.

Examples: Find all factors of 20.

$1 \times 20 = 20$ $2 \times 10 = 20$ $4 \times 5 = 20$
All the factors of 20 are: 1, 20, 2, 20, 4, 5

Find all factors of 84.

$1 \times 84 = 84$ $2 \times 42 = 84$ $3 \times 28 = 84$
$4 \times 21 = 84$ $6 \times 14 = 84$ $7 \times 12 = 84$

All the factors of 84 are: 1, 84, 2, 42, 3, 28, 4, 21, 6, 14, 7, 12

Find all the factors of each number.

S1. 30	S2. 36	1. 100	1.
			2.
2. 42	3. 70	4. 81	3.
			4.
5. 50	6. 40	7. 75	5.
			6.
8. 90	9. 20	10. 50	7.
			8.
			9.
			10.

Problem Solving

A test contained 60 questions. If a student's score was 90%, how many questions did he get correct?

Score

Review Exercises

1. $-9 - 6 + -3 =$

2. $-3 \times -2 \cdot 4 =$

3. $\sqrt{121} + \sqrt{81}$

4. Solve the proportion.

$$\frac{3}{4} = \frac{n}{10}$$

5. $3 = 20\%$ of what?

6. Two is what % of eight?

Helpful Hints

The **greatest common factor** is the largest factor that two or more numbers have in common.

Example: Find the greatest common factor of 12 and 16.

Find the factors of each number: 12: 1, 2, 3, ④, 6, 12
                              16: 1, 2, ④, 8, 16   greatest common factor = ④

* "Greatest common factor" is abbreviated as GCF.

Find the greatest common factor of each pair of numbers.

S1. 8 and 10   S2. 12 and 20   1. 6 and 8

2. 12 and 15   3. 42 and 56   4. 64 and 80

5. 100 and 120   6. 90 and 70   7. 45 and 25

8. 60 and 72   9. 48 and 36   10. 20 and 40

1.	
2.	
3.	
4.	
5.	
6.	
7.	
8.	
9.	
10.	
Score	

Problem Solving

Light travels approximately 5.879×10^{12} miles in one year.
Write the distance travelled as a conventional number.

Review Exercises

1. Write .0000012 in scientific notation.

2. Write 496,000,000 in scientific notation.

3. Write 1.32×10^7 as a conventional number.

4. Write $4.64 \times 10\text{-}6$ as a conventional number.

5. Find all the factors of 60.

6. Find the GCF (greatest common factor) of 36 and 40.

Helpful Hints

A **multiple** of a number is the product of that number and any whole number.

The multiples of a number can be found by multiplying it by 0, 1, 2, 3, 4, and so on.

Example: Find the first six multiples of 3.

3: 0, 3, 6, 9, 12, 15

These are found by multiplying 3 by 0, 1, 2, 3, 4, and 5.

Complete the list of multiples for each number.

S1. 2: 0, 2, ☐, ☐, ☐, ☐

S2. 6: ☐, 6, ☐, ☐, 24, ☐

1. 5: 0, 5, ☐, ☐, ☐, ☐

2. 3: ☐, 3, ☐, 9, ☐, ☐

3. 10: ☐, 10, 20, ☐, ☐, ☐

4. 4: ☐, ☐, ☐, 12, 16, 20

5. 11: 0, 11, ☐, 33, ☐, 55

6. 8: 0, 8, 16, ☐, ☐, ☐

7. 20: 0, 20, 40, ☐, ☐, ☐

8. 7: 0, 7, ☐, 21, ☐, ☐

9. 30: 0, 30, 60, ☐, ☐, ☐

10. 9: 0, 9, 18, ☐, 36, ☐

1.

2.

3.

4.

5.

6.

7.

8.

9.

10.

Problem Solving

A pitcher threw 30 pitches that were strikes.
This was 25% of all the pitches thrown.
How many pitches were thrown by the pitcher?

Score

Review Exercises

1. List all the factors of 30.

2. Find the GCF of 32 and 60.

3. List the first six multiples of eight.

4. Find 6% of 50.

5. 3 is what % of 12?

6. 7 = 20% of what?

Helpful Hints

The **least common multiple** of two or more whole numbers is the smallest whole number, other than zero, that they all divide into evenly.

Examples: The least common multiple of:
2 and 3 is 6   4 and 6 is 12   3 and 9 is 9

* Least common multiple is abbreviated as LCM.

Find the least common multiple of each pair of numbers.

S1. 3 and 4

S2. 6 and 8

1. 3 and 5

2. 6 and 10

3. 12 and 20

4. 10 and 15

5. 12 and 18

6. 15 and 60

7. 16 and 12

8. 8 and 20

9. 9 and 12

10. 12 and 30

1.

2.

3.

4.

5.

6.

7.

8.

9.

10.

Problem Solving

A CD costs $12. If the sales tax is 8%, what is the total cost of the CD?

Score

Reviewing Number Theory

For 1-6, find all factors for each number.

1. 24   2. 16   3. 32

4. 28   5. 70   6. 25

For 7-12, find the greatest common factor for each pair of numbers.

7. 12 and 8   8. 48 and 60   9. 120 and 100

10. 45 and 50   11. 35 and 28   12. 90 and 72

For 13 - 15, complete the list of multiples of each number.

13. 3: 0, 3, 6, ☐, ☐, ☐   14. 9: 0, ☐, 18, ☐, ☐, ☐,

15. 15: 0, ☐, ☐, ☐, ☐, 75

For 16-20, find the least common multiple of each pair of numbers.

16. 4 and 6   17. 12 and 15   18. 20 and 15

19. 12 and 4   20. 8 and 6

1.
2.
3.
4.
5.
6.
7.
8.
9.
10.
11.
12.
13.
14.
15.
16.
17.
18.
19.
20.

Review Exercises

1. Find the GCF of 40 and 56.

2. Find the LCM of 4 and 6

3. List all factors of 28.

4. List the first six multiples of 12.

5. Is $\frac{6}{8} = \frac{3}{4}$ a proportion? Why?

6. Solve the proportion.
$$\frac{6}{8} = \frac{n}{12}$$

Helpful Hints

Numbers can be assigned to a point on a **number line**. **Positive numbers** are to the right of zero. **Negative numbers** are to the left of zero.

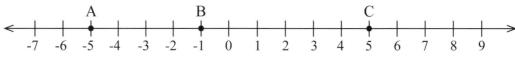

Numbers are graphed on a number line with a point.

Examples: A is the graph of -5.   A has a coordinate of -5.
              B is the graph of -1.   B has a coordinate of -1.
              C is the graph of 5.   C has a coordinate of 5.

Use the number line to state the coordinates of the given points.

S1. B

S2. D, E, and G

1. L and H

2. R and F

3. K, F, and C

4. N and A

5. G, H, I, and Q

6. H, D, and S

7. A, M, B, and P

8. B, C, and M

9. I, F, and P

10. L, P, H, and A

1.

2.

3.

4.

5.

6.

7.

8.

9.

10.

Problem Solving

At midnight the temperature was -4°.
By 6:00 a.m. the temperature had risen 12°.
What was the temperature at 6:00 a.m.?

Score

Review Exercises

Use A = {2,4,6,8,10}, B = {1,3,4,5,6,8,10}, and C = {4,5,6,8,9,10} to answer the following questions.

1. A ∩ B =   **2.** B ∪ C =   **3.** A ∪ C =   **4.** B ∩ C =

5. Are B and C equivalent sets? Why?   **6.** Are A and C disjoint sets? Why?

Helpful Hints

Equations can be solved and graphed on a **number line**.

Examples:

x + 5 = 7	n - 3 = 2	3y = 21	$\frac{m}{2} = 5$
2 + 5 = 7	5 - 3 = 2	3 × 7 = 21	$\frac{10}{2} = 5$
x = 2	x = 5	y = 7	
			m = 10

Solve each equation and graph each solution on the number line.
Also place each solution in the answer column.

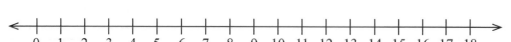

S1. x + 2 = 3   **S2.** y - 2 = 5   **1.** c + 4 = 7

2. 5 - e = 0   **3.** 3d = 15   **4.** $\frac{f}{3} = 6$

5. n × 2 = 8   **6.** 3 + j = 14   **7.** 3 + k = 11

8. 4m = 28   **9.** 6 = n + 2   **10.** $\frac{r}{2} = 6$

1.
2.
3.
4.
5.
6.
7.
8.
9.
10.
Score

Problem Solving

A car can travel 320 miles in five hours.
At this rate, how far can it travel in eight hours?

Review Exercises

1. -2 + 9 =

2. -7 - 15 =

3. -7 - -15 =

4. 6 × -7 =

5. -45 ÷ -9 =

6. $\dfrac{-24 \div -2}{18 \div -3}$ =

Helpful Hints	**Ordered pairs** can be graphed on a **coordinate system**. The first number of an ordered pair shows how to move across. It is called the **x-coordinate**. The second number of an ordered pair shows how to move up and down. It is called the **y-coordinate**. Examples: To locate B, move across to the right to 3 and up 4. The ordered pair is (3,4). To locate C, move across to the left to -5 and up 2. The ordered pair is (-5,2).	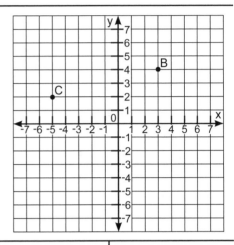

Use the coordinate system to find the point associated with each ordered pair.

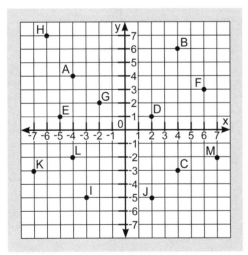

S1. D

S2. L

1. F

2. J

3. K

4. E

5. B

6. C

7. I

8. G

9. D

10. H

1.

2.

3.

4.

5.

6.

7.

8.

9.

10.

Problem Solving	A shirt that regularly sells for $30 is on sale for 20% off. How much is the sale price?

Score

Review Exercises

1. $\frac{1}{3} + -\frac{4}{5} =$

2. $-.29 + -.39 =$

3. $-\frac{1}{8} - (-\frac{1}{2}) =$

4. $-\frac{2}{3} \times -1\frac{1}{2} =$

5. $2\frac{1}{2} \div -\frac{1}{2} =$

6. $-5 \div -2\frac{1}{2} =$

Helpful Hints

A **point** can be found by matching it with an ordered pair.

Examples: (-5, 3) is found by moving across to the left to -5, and up 3. This is represented by point B. -5 is the **x-coordinate** and 3 is the **y-coordinate**.

(6, 3) is found by moving across to the right to 6, and up 3. This is represented by point C. 6 is the **x-coordinate** and 3 is the **y-coordinate**.

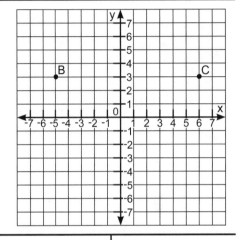

Use the coordinate system to find the point associated with each ordered pair.

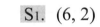

S1. (6, 2)   S2. (-5, 5)

1. (3, 5)   2. (7, -6)

3. (-6, -4)   4. (0, 3)

5. (-2, -4)   6. (-2, 2)

7. (-6, 2)   8. (4, 2)

9. (-4, -7)   10. (4, -3)

1.

2.

3.

4.

5.

6.

7.

8.

9.

10.

Score

Problem Solving

In a class of 40 students, 38 were present. What percent of the class was present?

Review Exercises

1. $2^5 =$

2. $\sqrt{36} + 4^2 =$

3. $\dfrac{4^2 + 3^2}{\sqrt{25}} =$

4. $2^3 \times 3^2 =$

5. Write .00017 in scientific notation.

6. Write 213,000 in scientific notation.

Helpful Hints

The **slope** of a line refers to how steep the line is. It is the ratio of **rise to run**.

$$slope = \frac{y_2 - y_1}{x_2 - x_1}$$

Example:
What is the slope of the line passing through the ordered pairs (1, 5) and (6, 9)?

$$slope = \frac{y_2 - y_1}{x_2 - x_1} \qquad \begin{matrix} x_1 \ y_1 & x_2 \ y_2 \\ (1, 5), & (6, 9) \end{matrix}$$

$$= \frac{9 - 5}{6 - 1}$$

$$= \boxed{\frac{4}{5}} \qquad \text{The slope is } \frac{4}{5}$$

The run is 5 and the rise is 4.

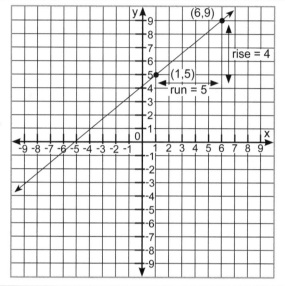

Find the slope of each line that passes through the given point.

S1. (2, 3), (5, 4)   S2. (3, -2), (5, 1)   1. (4, 3), (2, 6)

2. (4, 1), (7, 2)   3. (-2, 1), (-3, 3)   4. (-2, -2), (6, 3)

5. (4, 5), (6, 6)   6. (1, 2), (3, 9)   7. (1, -1), (6, 5)

8. (3, 2), (8, 6)   9. (2, -1), (4, 2)   10. (9, 2), (7, 5)

1.

2.

3.

4.

5.

6.

7.

8.

9.

10.

Problem Solving

In a school the ratio of boys to girls is five to four. If there are 400 boys, how many girls are there in the school?

Score

Reviewing Number Lines and Coordinate Systems

Use the number line to state the coordinates of the given points.

1. C   2. B, F, and J   3. S, M, and N   4. R, T, C, and D

Solve each equation and graph each solution on the number line. Be sure to label your answers. Also, place each solution in the answer column.

5. $n + 2 = 5$ 6. $x - 2 = 4$ 7. $3y = 15$ 8. $\frac{m}{2} = 4$

Use the coordinate system to find the ordered pair associated with each point.

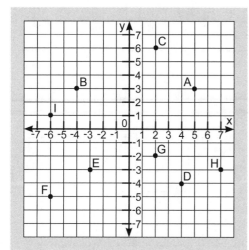

9. A

10. I

11. D

12. F

13. C

14. Find the slope of the line that passes through points A and G.

Use the coordinate system to find the ordered pair associated with each point.

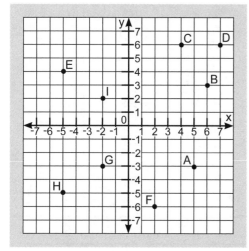

15. (6, 3)

16. (-2, 2)

17. (-5, -5)

18. (7, 6)

19. (5, -3)

20. Find the slope of the line that passes through points B and I.

1.
2.
3.
4.
5.
6.
7.
8.
9.
10.
11.
12.
13.
14.
15.
16.
17.
18.
19.
20.

47

Review Exercises

1. $-36 \div -6 =$

2. $-9 -6 + -3 =$

3. $-2 \times -3 \times -4 =$

4. $-7 - 9 =$

5. $-56 \div 8 =$

6. $(-2)^3 =$

Helpful Hints

The graph of a **linear equation** is always a line. A linear equation can have an infinite number of solutions, so to make a graph we select a few points and graph them, and then draw a line that connects them.

Example: Draw a graph of the solutions to the following equation.

$$y = x + 3$$

First, select four values for x and find the values for y. Start with x = 0 and make a chart like the one to the right.

x	y	
0	3	(0,3)
1	4	(1,4)
2	5	(2,5)
4	7	(4,7)

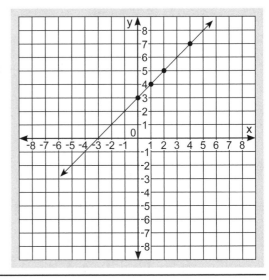

Next, plot the points and connect them with a line.

Make a table of 4 solutions. Graph the points. Connect them with a line

S1.

$y = 2x + 1$

x	y

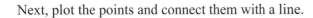

S2.

$y = x - 3$

x	y

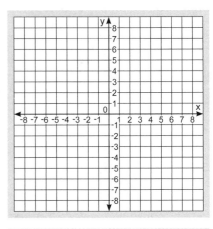

1.

$y = 2x - 1$

x	y

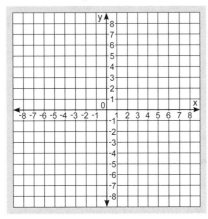

2.

$y = \dfrac{x}{2}$

x	y

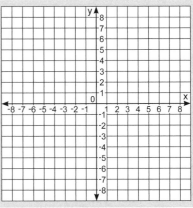

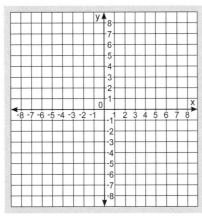

Problem Solving   If three pounds of meat costs $3.60, how much will five pounds costs?

Review Exercises

1. Solve the proportion.
$$\frac{5}{6} = \frac{7}{n}$$

2. Find 15% of 20.

3. 15 = 20% of what?

4. $-\frac{1}{3} + -\frac{3}{8} =$

5. $2 \times -1\frac{1}{2} =$

6. $-6.3 \div 3 =$

Helpful Hints

Use what you have learned to work the following problems.

Example: Draw a graph of the solutions to the following equation.

$$y = \frac{x}{2} + 2$$

x	y	
0	2	(0,2)
2	3	(2,3)
4	4	(4,4)
-2	1	(-2,1)

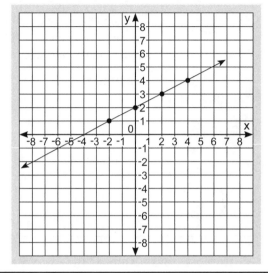

Make a table of 4 solutions. Graph the points. Connect them with a line

S1.

$y = \frac{x}{3}$

x	y

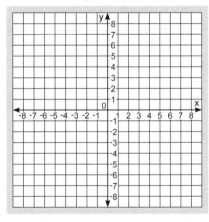

S2.

$y = 2x + 3$

x	y

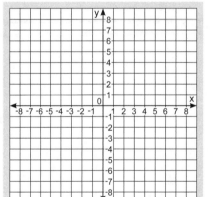

1.

$y = \frac{x}{2} + 5$

x	y

2.

$y = -2x$

x	y

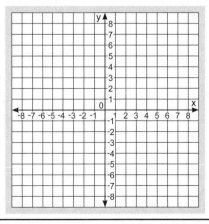

Problem Solving   John has finished $\frac{4}{5}$ of the problems on a test. What percent has he finished?

Review Exercises

1. Write 1,720,000 in scientific notation.

2. Write .00000038 in scientific notation.

3. Write 1.963×10^8 as a conventional number.

4. Write 3.4×10^{-4} as a conventional number.

5. -9 - 7 - -6 =

6. -.34 + .53 =

Helpful Hints

The goal with any **equation** is to end up with the **variable** (letter), an **equal sign**, and the **answer**. You can add, subtract, multiply, and divide on each side of the equal sign with the same number, and won't change the solution.

Examples:

$$x + 2 = 9$$
$$\underline{+\ -2 = -2}$$
$$x = \boxed{7}$$
Add -2 to both sides.

$$n - 6 = -5$$
$$\underline{+\ 6 = 6}$$
$$n = \boxed{1}$$
Add 6 to both sides.

$$4n = 24$$
$$\frac{4n}{4} = \frac{24}{4}$$
$$n = \boxed{6}$$
Divide both sides by 4.

$$\frac{x}{6} = 4$$
$$\frac{6}{1} \times \frac{x}{6} = 4 \times 6$$
$$x = \boxed{24}$$
Multiply both sides by 6.

Check your work by substituting your answer in the original equation.

Solve the equations. Refer to the examples above.

S1. $x + 3 = 8$

S2. $3n = 96$

1. $n - 5 = -8$

2. $\frac{n}{5} = 8$

3. $n + 6 = -7$

4. $5n = -25$

5. $n + -6 = 7$

6. $\frac{n}{4} = -3$

7. $x + 23 = 57$

8. $15n = 60$

9. $n - -6 = -5$

10. $n + 12 = -15$

1.
2.
3.
4.
5.
6.
7.
8.
9.
10.

Problem Solving

What is the slope of a line that passes through the points (6, 1) and (9, 8)?

Score

Review Exercises

1. $\dfrac{3^2 + 2^2 + 7}{2} =$

2. $3 + 7 \times 2 + 6 =$

3. $3 \times (7^2 - 15) =$

4. $n + 5 = -3$

5. $3n + 18 =$

6. $\dfrac{n}{3} = 7$

Helpful Hints

Be careful with negative signs when solving equations.

Examples:

$$\begin{array}{rr} -x + 7 = & -9 \\ + \quad -7 = & -7 \\ \hline -x = & -16 \end{array}$$

If -x = -16, then x = $\boxed{16}$

$$\begin{array}{l} -3n = 18 \\ \dfrac{-3n}{-3} = \dfrac{18}{-3} \end{array}$$

Divide both sides by -3.

n = (-6)

$$\dfrac{n}{-5} = 7$$

$$\dfrac{-5}{1} \times \dfrac{n}{-5} = 7 \times -5$$

Multiply both sides by -5.

n = (-35)

* Remember to check your work by substituting your answer in the original equation.

Solve the equations. Refer to the examples above.

S1. $-x + 7 = -5$

S2. $-4n = -12$

1. $-n - 6 = 8$

2. $\dfrac{n}{-2} = 6$

3. $-x + -7 = 2$

4. $-3n = 15$

5. $n - 6 = 12$

6. $5n = -30$

7. $-n - 6 = -8$

8. $\dfrac{n}{-4} = -5$

9. $3n = -45$

10. $-n + -6 = -20$

1.	
2.	
3.	
4.	
5.	
6.	
7.	
8.	
9.	
10.	
Score	

Problem Solving

Write the ratio 18 to 8 as a fraction reduced to lowest terms.

Review Exercises

1. -7 - -9 + 6 - 7 =

2. 3 × -2 × 4 × -3 =

3. $\dfrac{-64 \div 8}{24 \div -6}$ =

4. Write 210,000 in scientific notation.

5. Write .00316 in scientific notation.

6. $(-2)^4$ =

Helpful Hints

Some equations require two steps.

Examples:

$$\begin{array}{ll} 2x - 5 = 71 & \text{Add 5 to} \\ +\ 5 = \ 5 & \text{both sides.} \\ \hline \dfrac{2x}{2} = \dfrac{76}{2} & \text{Divide both} \\ & \text{sides by 2.} \\ x = \boxed{38} \end{array}$$

$$\begin{array}{ll} \dfrac{n}{5} + 3 = 8 & \text{Add 3 to} \\ +\ -3 = -3 & \text{both sides.} \\ \hline \dfrac{5}{1} \times \dfrac{n}{5} = 5 \times 5 & \text{Multiply both} \\ & \text{sides by 5.} \\ n = \boxed{25} \end{array}$$

$$\begin{array}{ll} -3n - 4 = 11 & \text{Add 4 to} \\ +\ 4 = \ 4 & \text{both sides.} \\ \hline \dfrac{-3n}{-3} = \dfrac{15}{-3} & \text{Divide both} \\ & \text{sides by -3.} \\ n = \boxed{-5} \end{array}$$

* Remember to check your work by substituting your answer in the original equation.

S1. $3x - 5 = 16$

S2. $\dfrac{x}{2} + 2 = 4$

1. $7x + 3 = -4$

2. $-14n - 7 = 49$

3. $2n + 45 = 15$

4. $\dfrac{n}{5} + -6 = 9$

5. $4x - 10 = 38$

6. $-2m + 9 = 7$

7. $35x + 12 = 82$

8. $\dfrac{m}{5} - 7 = 3$

9. $3x - 12 = 18$

10. $5x + 2 = -13$

1.
2.
3.
4.
5.
6.
7.
8.
9.
10.
Score

Problem Solving

Six students were absent Monday at Jefferson School. If this was 3% of the total enrollment, how many students are enrolled at Jefferson School?

Review Exercises

1. $-\dfrac{2}{5} + \dfrac{1}{2} =$

2. $-\dfrac{2}{5} + -\dfrac{2}{5} =$

3. $\dfrac{2}{3} \div -\dfrac{1}{2} =$

4. $1\dfrac{1}{2} \times -2 =$

5. $.2 \times -3.2 =$

6. $-.6 + -.5 =$

Helpful Hints

Sometimes the **distributive property** can be used to solve equations.

Examples:

$2(x + 7) = 30$
First use the distributive property.

$2(x + 7) = 30$
$2x + 14 = 30$
$\underline{\quad + -14 = -14}\quad$ Add -14 to both sides.
$\dfrac{2x}{2} = \dfrac{16}{2}\quad$ Divide both sides by 2.
$x = \boxed{8}$

$3(4x - 3) = -33$
First use the distributive property.

$3(4x - 3) = -33$
$12x - 9 = -33$
$\underline{\quad + 9 = 9}\quad$ Add 9 to both sides.
$\dfrac{12x}{12} = \dfrac{-24}{12}\quad$ Divide both sides by 12.
$x = \boxed{-2}$

* Remember to check your answers.

Solve the following equations. Use the distributive property when necessary.

S1. $5(m + 6) = 45$

S2. $\dfrac{x}{5} + -6 = 3$

1. $3(m - 2) = 18$

2. $3x + 7 = -2$

3. $4m - 9 = 31$

4. $2(m + -2) = -10$

5. $-6x + 2 = -28$

6. $-x + 8 = 12$

7. $\dfrac{x}{2} + 3 = -2$

8. $2x + 1 = -13$

9. $5x - 3 = -18$

10. $4(x + 2) = 48$

1.	
2.	
3.	
4.	
5.	
6.	
7.	
8.	
9.	
10.	
Score	

Problem Solving

Find the greatest common factor of 42 and 56.

Review Exercises

1. $6^3 =$

2. $7^0 =$

3. $9^1 =$

4. $\sqrt{36} + \sqrt{49} =$

5. $2^3 + \sqrt{16}$

6. $33 + 5^3 =$

Helpful Hints

Sometimes there are variables on both sides of the equal sign.

Examples:

$$
\begin{array}{ll}
5x - 6 = 2x + 9 & \\
\underline{+\ -2x\ \ \ \ = -2x} & \text{Add -2x to both sides.} \\
3x - 6 = 9 & \\
\underline{+\ \ \ 6 = 6} & \\
\dfrac{3x}{3} = \dfrac{3}{3} & \text{Divide both sides by 3.} \\
x = \boxed{1} &
\end{array}
$$

$$
\begin{array}{ll}
-6x + 12 = 4x - 8 & \\
\underline{+6x\ \ \ \ \ \ = 6x} & \text{Add 6x to both sides.} \\
12 = 10x - 8 & \\
\underline{+\ \ 8 = \ \ \ \ \ \ 8} & \\
\dfrac{20}{10} = \dfrac{10x}{10} & \text{Divide both sides by 10.} \\
\boxed{2} = x &
\end{array}
$$

S1. $4x + 3 = 2x + 1$

S2. $5x + 1 = 2x + 10$

1. $4x - 12 = 2x + 2$

2. $x - 2 = 2x - 4$

3. $7x - 16 = x + 8$

4. $7x - 1 = 15 + 3x$

5. $-2x + 8 = 4x - 10$

6. $3x + 6 = x + 8$

7. $3x - 5 = x - 7$

8. $3x + 5 = x + 13$

9. $2x + 6 = -x + 12$

10. $5x - 6 = 2x + 12$

1.	
2.	
3.	
4.	
5.	
6.	
7.	
8.	
9.	
10.	
Score	

Problem Solving

Susan had 30 apples and used six of them to make a pie. What percent of the apples did she use to make the pie?

Review Exercises

1. Solve the proportion.
$$\frac{n}{4} = \frac{25}{5}$$

2. Is $\frac{4}{7} = \frac{3}{5}$ a proportion? Why?

3. Write 16 to 6 as a fraction reduced to lowest terms.

4. Find 20% of 300.

5. Six is what % of 24?

6. $7 = 20\%$ of what?

Helpful Hints

Use what you have learned to solve the following equations.
* If necessary, refer to the previous Helpful Hints sections.
* Check your answers by substituting them in the original equation.

Solve the following equations. Use the distributive property when necessary.

S1. $7n = 28$

S2. $\frac{n}{5} = 3$

1. $2x + 6 = 2$

2. $3x + 7 = -2$

3. $4m - 9 = 31$

4. $4(x + 3) = -8$

5. $3(n - 2) = 30$

6. $3x + 6 = x + 8$

7. $4x - 12 = 2x + 2$

8. $-5x + 2 = -13$

9. $\frac{x}{5} - 2 = -7$

10. $3(x + 4) = -18$

1.	
2.	
3.	
4.	
5.	
6.	
7.	
8.	
9.	
10.	
Score	

Problem Solving

Jeff has a marble collection. The ratio of red marbles to blue marbles is three to two. If he has 12 red marbles, how many blue marbles does he have? (Use a proportion.)

Review Exercises

1. List all the factors of 48.

2. What is the GCF of 16 and 24?

3. What is the LCM of 6 and 10?

4. $3n = 15$, n =

5. $\frac{n}{2} = 10$, n =

6. $-x = 5$, x =

Helpful Hints

To solve **algebra word problems**, it is necessary to translate words into **algebraic expressions** containing a **variable**. A **variable** is a letter that represents a number. Here are some examples:

Three more than a number → x + 3
Twice a number → 2x
The quotient of x and five → $\frac{x}{5}$
Seven less than three times a number → 3x - 7
Twice a number less nine is equal to 15 → 2x - 9 = 15
The difference between three times a number and eight equals 12 → 3x - 8 = 12
The sum of a number and -9 is 24 → x + -9 = 24
Three times a number less six equals twice the number plus 15 → 3x - 6 = 2x + 15
Twice the sum of n and five → 2(n + 5)
The difference between four times x and 15 equals twice the number → 4x - 15 = 2x

Four less than a number → x - 4
Seven times a number → 7x
A number decreased by six → x - 6

Translate each of the following into an equation.

S1. Seven less than twice a number is 12.

S2. Two more than three times a number equals 30.

1. The sum of twice a number and five is 14.

2. The difference between four times a number and six is 10.

3. Twelve is five less than four times a number.

4. One-third times a number less four equals twice the number added to eight.

5. Twice the sum of a number and two equals 10.

6. The difference between five times a number and three is 17.

7. Twice a number decreased by six is 15.

8. Two less than three times a number is seven more than twice the number.

9. Four more than a number equals the sum of seven and -12.

10. A number divided by five is 25.

| 1. |
| 2. |
| 3. |
| 4. |
| 5. |
| 6. |
| 7. |
| 8. |
| 9. |
| 10. |
| Score |

Problem Solving

If a car can travel 65 miles per hour, how far can it travel in 3.5 hours?

Review Exercises

1. $x + 2 = 9$
   $x =$

2. $n + -3 = -7$
   $n =$

3. $3n = 36$

4. $-5n = -25$

5. $\frac{n}{3} = 5$
   $n =$

6. $2x + 1 = 7$

Helpful Hints

Algebra word problems must be translated into an **equation** and solved.

Example:

Six times a number less two equals four times the number added to 10.
First translate and then solve.

$$6x - 2 = 4x + 10$$

$$\underline{+ -4x \qquad -4x}$$ Add -4x to both sides.

$$2x - 2 = 10$$

$$2 = \ 2$$ Add 2 to both sides.

$$\frac{2x}{2} = \frac{12}{2}$$ Divide both sides by 2.

$$x = \boxed{6}$$ The number is 6.

Translate each of the following into an equation and solve.

S1. Six less than twice a number is 16. Find the number.

S2. The difference between three times a number and 8 is 28.
    Find the number.

1. Five less than twice a number is 67. Find the number.

2. Four times a number decreased by five is -17. Find the number.

3. Four times a number less six is eight more than two times the number.
   Find the number.

4. Eight more then one-half a number is 10. Find the number.

5. The difference between four times a number and two is 10.

1.
2.
3.
4.
5.
Score

Problem Solving

A doctor's annual income is $150,000. What is his average monthly income?

Review Exercises

1. Write 3.61×10^{-7} as a conventional number.

2. Write .00000127 in scientific notation.

3. Write 729,000,000 in scientific notation.

Helpful Hints

Remember these steps when solving algebra word problems.

1. Read the problem very carefully.
2. Write an equation.
3. Solve the equation and find the answer.
4. Check your answer to be sure it makes sense.

Example: John is twice as old as Susan. The sum of their ages is 42. What is each of their ages?

Let x = Susan's age   2x = John's age

$x + 2x = 42$ Susan's age is x = 14.
$3x = 42$ John's age is 2x = 28.
$x = 14$ The sum is 42.

Solve the algebra word problems.

S1. Amir is six years older than Kevin. The sum of their ages is 30. Find the age of each.

S2. A board 44 inches long is cut into two pieces. The long piece is three times the length of the short piece. What is the length of each piece.

1. Bob and Bill together earn $66. Bill earned $6 more than twice as much as Bob. How much did each earn?

2. Steve worked Monday and Tuesday and earned a total of $212. He earned $30 more on Tuesday than he did on Monday. How much did Steve earn each day?

3. Five times Bob's age plus six equals three times his age plus 30. What is Bob's age?

4. Sixty dollars less than three times Susan's weekly salary is equal to 360 dollars. What is Susan's weekly salary?

5. Twice John's age less 12 is 48. What is John's age?

1.
2.
3.
4.
5.
Score

Problem Solving

A student has test scores of 90, 96, 84, and 86. What was his average score?

Review Exercises

Solve each equation.

1. $2x + 7 = -15$

2. $5x + 6 = 106$

3. $\dfrac{n}{4} + 2 = 13$

4. $3(n + 6) = -9$

5. $5x + 3 = 7x + -3$

6. $3x + 2x = 55$

Helpful Hints

*Remember:
1. Read the problem carefully.
2. Write an equation.
3. Solve the equation and find the answer.
4. Check your answer to be sure it makes sense.

Solve each algebra word problem.

S1. Five more than six times a number is equal to 48 less than seven. Find the number.

S2. Steve weighs 50 pounds more than Bart. Their combined weight is 270 pounds. What is each of their weights?

1. The sum of three times a number and 15 is -12. Find the number.

2. Eight more than six times a number is 20 more than four times the number. Find the number.

3. The sum of five and a number is -19. Find the number.

4. Roy is three times as old as Ellen. The sum of their ages is 44 years. What are each of their ages?

5. Six more than two times a number is six less than six times the number. Find the number.

1.	
2.	
3.	
4.	
5.	
Score	

Problem Solving

A plane travelled 2,100 miles in 3.5 hours.
What was the plane's average speed per hour?

Reviewing Equations and Algebra Word Problems

For 1 - 12, solve each equation. Be sure to show all work.

1. $x + 5 = -2$	2. $3n = 39$	3. $\frac{n}{7} = 8$
4. $5n + 2 = 17$	5. $3n - 6 = -21$	6. $\frac{n}{3} - 6 = -12$
7. $3(n + 2) = -15$	8. $5(x - 4) = 55$	9. $2x + 4 = 4x - 12$
10. $5x - 3 = 3x + 13$	11. $3x + 4x = -77$	12. $\frac{n}{-3} + 2 = -5$

For 13 - 20, solve each algebra word problem

13. Twice a number less three is 21.
Find the number.

14. Eight more than five times a
number is -17. Find the number.

15. The difference between five times
a number and six is 24. Find
the number.

16. Seven more than twice a number
is five less than four times the
number. Find the number.

17. Ann has twice as much money as
Sue. Together they have $66.
How much does each have?

18. Bill is eight years older than Ron.
The sum of their ages is 64 years.
How old is each of them?

19. Four times a number decreased
by six equals -14. Find the number.

20. Four more than one-third of a
number is 10. Find the number.

1.
2.
3.
4.
5.
6.
7.
8.
9.
10.
11.
12.
13.
14.
15.
16.
17.
18.
19.
20.

Review Exercises

1. List the first seven multiples of 8.

2. List all factors of 60.

3. What is the GCF of 100 and 40?

4. Write .000006 in scientific notation.

5. Write 2,100,000 in scientific notation.

6. Write 2.1×10^{-3} as a conventional number.

Helpful Hints

Probability tells what chance, or how likely it is for an event to occur. Probability can be written as a fraction.

$$\text{Probability} = \frac{\text{number of ways a certain outcome can occur}}{\text{number of possible outcomes}}$$

Examples: If you toss a coin, what is the probability that it will show heads?

$\frac{1 - \text{heads is one outcome}}{2 - \text{there are two possible outcomes, heads or tails}}$ The probability is 1 out of 2.

There are six marbles in a jar. Three are red, two are blue, and one is green. What is the probability that you will draw a blue one without looking?

$\frac{2 - \text{blue marbles}}{6 - \text{marbles in the jar}}$ The probability is 2 out of 6, or simplified, 1 out of 3.

Use the information below to answer the following questions.

There are 3 red marbles, 6 blue marbles, 2 black marbles, and 1 green marble in a can. Find the probability of each of the following.

S1. A red marble.

S2. A blue or green marble.

1. A black marble.

2. A green marble.

3. A blue or red marble.

4. Not a black marble.

5. Not a red marble.

6. Not a green or blue marble.

7. A green, red, or blue marble.

8. Not a blue marble.

9. A green, red, or black marble.

10. Not a blue or black marble.

1.
2.
3.
4.
5.
6.
7.
8.
9.
10.
Score

Problem Solving

Four times a number less five is -17. Find the number.

Review Exercises

Solve each of the following equations.

1. $3x + 2 = -28$ 2. $\frac{x}{5} - 6 = -11$ 3. $4(n + 3) = -28$

4. $2x + 10 = 4x + 2$ 5. $3x + 2x = 75$ 6. $7x - 3 = 60$

Helpful Hints

Use what you have learned to solve the following questions.

Example: What is the probability of the spinner landing on the 1 or the 3?

2 out of 8 or, simplified, 1 out of 4.

Use the spinner to find the probability for each of the following.
Find the probability of spinning once and landing on each of the following.

S1. a three   S2. an even number

1. a seven   2. not a five

3. an odd number   4. a number less than five   5. a number greater than six

6. a nine   7. a one or an eight   8. an even number or a five

9. a number greater than three   10. a number which is a factor of six

1.	
2.	
3.	
4.	
5.	
6.	
7.	
8.	
9.	
10.	
Score	

Problem Solving

If five pounds of beef cost $9, how many pounds can be bought with $36?

Review Exercises

1. Change .3 to a percent.
2. Change .03 to a percent.
3. Change $\frac{3}{5}$ to a percent.

4. Find 4% of 50.
5. Fifteen is what % of 60?
6. 4 = 20% of what?

Helpful Hints

Statistics involves gathering and recording **data**. Number facts about events or objects are called data. The **range** is the difference between the greatest number and the least number in a list of data. The **mode** is the number which appears the most in a list of data.

Example: Find the range and mode for the list of data.
12, 10, 1, 7, 4, 7, 5
First, list the numbers from least to greatest.
1, 4, 5, 7, 7, 10, 12
The range is 12 - 1 = 11.
The mode is 7, which appears the most.

Arrange the data in order from least to greatest, then find the range and mode.

S1. 7, 4, 1, 8, 2, 5, 4

S2. 6, 2, 7, 6, 8, 2, 5, 6, 3

1. 7, 4, 8, 2, 4, 7, 7

2. 25, 17, 30, 39, 16, 24, 30

3. 1, 3, 6, 3, 4, 6, 11, 9

4. 1, 6, 17, 8, 9, 20, 9

5. 7, 3, 1, 3, 1, 3, 8, 4

6. 3, 14, 8, 6, 11, 8, 14, 8

7. 1, 10, 2, 9, 3, 8, 2, 7

8. 85, 91, 90, 86, 91, 87

9. 1, 10, 2 9, 2, 7, 2, 8

10. 20, 2, 19, 1, 2, 16, 3

1.	
2.	
3.	
4.	
5.	
6.	
7.	
8.	
9.	
10.	
Score	

Problem Solving

If three cans of juice cost $1.14, what is the cost of one can?

Review Exercises

1. Write 16 to 10 as a fraction reduced to lowest terms.

2. Is $\frac{9}{11} = \frac{7}{8}$ a proportion? Why?

3. Solve the proportion.
   $$\frac{4}{n} = \frac{9}{45}$$

4. Write 1,280,000 in scientific notation.

5. Write .0000962 in scientific notation.

6. Write 6.2×10^{-5} as a conventional number.

Helpful Hints

The **mean** of a list of data is found by adding all the items in the list and then dividing by the number of items.

The **median** is the middle number, when the list of data is arranged from least to greatest.

Example: Find the mean and median for the list of data.

1, 2, 5, 6, 6

Median = ⑤   Mean = $\dfrac{1 + 2 + 5 + 6 + 6}{5} = \dfrac{20}{5} = $ ④

Arrange the data in order from least to greatest, then find the mean and median.

S1. 1, 5, 2, 4, 3

S2. 6, 1, 7, 4, 2, 6, 2

1. 2, 7, 1, 4, 1

2. 1, 5, 7, 1, 2, 2, 3

3. 5, 25, 10, 20, 15

4. 1, 1, 1, 3, 3, 3, 4, 1, 1

5. 8, 5, 2, 9, 3, 6, 9

6. 126, 136, 110

7. 7, 3, 4, 2, 4

8. 3, 1, 4, 7, 5

9. 2, 10, 4, 8, 1

10. 50, 70, 30

1.	
2.	
3.	
4.	
5.	
6.	
7.	
8.	
9.	
10.	
Score	

Problem Solving

In a class of 40 students, 20% of them received A's. How many students did not receive A's?

Review Exercises

1. $3 + 4 \times 5 - 2 =$ 2. $3(8 + 2) - 4^2 =$ 3. $(15 - 8) + 64 \div 2^3 =$

4. $7 \times 4 - 9 \div 13 =$ 5. $6 [(3 + 4) \times 2] =$ 6. $3(-2 + 4) + 5 =$

Helpful Hints

Use what you have learned to answer the following questions.

* If necessary, refer to the two previous pages.

Arrange the data in order from least to greatest, then answer the questions.

2, 8, 6, 2, 7

S1. What is the range?   S2. What is the mode?

1. What is the mean?   2. What is the median?

1, 9, 2, 7, 2, 3, 4

3. What is the median?   4. What is the mode?

5. What is the range?   6. What is the mean?

2, 11, 8, 6, 1, 2, 5

7. What is the range?   8. What is the mode?

9. What is the mean?   10. What is the median?

1.	
2.	
3.	
4.	
5.	
6.	
7.	
8.	
9.	
10.	
Score	

Problem Solving

Light travels at a speed of 1.86×10^5 miles per second.
Write the speed as a conventional number.

Reviewing Probability and Statistics

There are four green marbles, three red marbles, two white marbles, and one blue marble in a can. What is the probability for each of the following?

1. a red marble

2. a green marble

3. a green or blue marble

4. not a red marble

5. a green, red, or blue marble

6. not a green marble

Use the spinner to find the probability of spinning once and landing on each of the following.

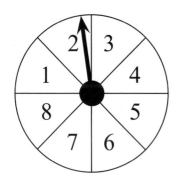

7. a five

8. an odd number

9. a number greater than three

10. a one or a three

11. a number less than five

12. a one or a six

Arrange the data in order from least to greatest, then answer the questions.

2, 5, 4, 10, 4

13. What is the range?

14. What is the mode?

15. What is the mean?

16. What is the median?

2, 5, 2, 1, 3, 7, 8

17. What is the mode?

18. What is the mean?

19. What is the range?

20. What is the median?

1.
2.
3.
4.
5.
6.
7.
8.
9.
10.
11.
12.
13.
14.
15.
16.
17.
18.
19.
20.

Final Review - All Pre-Algebra Concepts

For 1 - 3, use the following sets to find the answers.

$$A = \{1,2,3,4,5\}, \quad B = \{2,3,4,6,8\}, \quad C = \{0,1,2,4,5,9\}$$

1. Find $A \cap B$

2. Find $B \cup C$

3. Find $A \cap C$

4. $-9 + 12 =$

5. $-16 - 7 =$

6. $-12 \times -3 =$

7. $-24 \div -3 =$

8. $.21 + -.76 =$

9. $-\frac{2}{5} + -\frac{1}{2} =$

10. $5^3 =$

11. $\sqrt{49}$

12. $3^3 + \sqrt{36} =$

13. $6 + 7 \times 3 - 5 =$

14. $3^2(3 + 4) + 5 =$

15. $\dfrac{4^2 + 12}{5 + 3(2+1)} =$

16. $2[(5 + 7) \div 3 + 6] =$

17. What property is illustrated below?

   $5 + 6 = 6 + 5$

18. What property is illustrated below?

   $7(6 + 5) = 7(6) + 7(5)$

19. Write 1,280,000,000 in scientific notation.

20. Write .00000653 in scientific notation.

1.
2.
3.
4.
5.
6.
7.
8.
9.
10.
11.
12.
13.
14.
15.
16.
17.
18.
19.
20.

Final Review - All Pre-Algebra Concepts

21. Write 6.09×10^7 as a conventional number.

22. Write $7.62 \times 10\text{-}6$ as a conventional number.

23. Write 18 to 10 as a fraction reduced to lowest terms.

24. Is $\frac{7}{8} = \frac{3}{4}$ a proportion? Why?

25. Solve the proportion.
$$\frac{7}{n} = \frac{3}{9}$$

26. The ratio of red marbles to blue marbles is five to two. If there are 15 red marbles, how many blue marbles are there?

27. Find 5% of 80.

28. Six is what percent of 24?

29. $8 = 25\%$ of what?

30. On a test with 30 questions, a student got 80% correct. How many questions did he get correct?

31. There are 35 fish in an aquarium. If 14 of them are goldfish, what percent of them are goldfish?

32. Six students get A's on a test. This is 20% of the class. How many are there in the class?

33. Find all the factors of 40.

34. Find the GCF of 60 and 40.

35. Find the first seven multiples of six.

36. Find the LCM of 8 and 12.

Use the number line to state the coordinates of the given points.

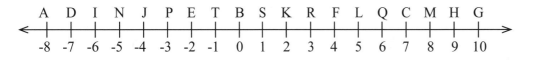

37. A, E, B   38. N, H, T, D   39. I, F, P   40. H, M, S

21.
22.
23.
24.
25.
26.
27.
28.
29.
30.
31.
32.
33.
34.
35.
36.
37.
38.
39.
40.

For 41 - 50, use the coordinate system to answer each question.

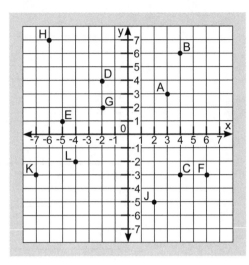

For 41-45, find the ordered pair associated with each point.

41.	F	42.	E

43.	C	44.	L

45.   B

For 46-52, find the point associated with each ordered pair.

46. (-6, 7)   47. (-2, 2)   48. (-4, -2)   49. (2, -5)   50. (-7, -3)

51. Find the slope of the line that passes through the points (1, 3) and (4, 5).

52. Find the slope of the line that passes through the points (-2, 5) and (6, 8).

For 53 through 54, make a table of 4 solutions and graph the points. Connect them with a line.

53.
$$y = x + 4$$

x	y

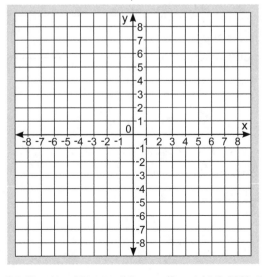

54.
$$y = 2x + 2$$

x	y

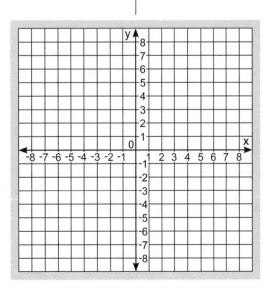

41.
42.
43.
44.
45.
46.
47.
48.
49.
50.
51.
52.
53.
54.

Solve each equation and word problem.

55. $x + 3 = 12$

56. $3n = -45$

57. $\dfrac{n}{6} = 3$

58. $-5n = 15$

59. $2x + 3 = 15$

60. $5x - 2 = -17$

61. $\dfrac{n}{3} + -4 = 4$

62. $3(x + 4) = 24$

63. $3(x + 4) = -6$

64. $2x + 12 = 4x + 10$

65. Two more than three times a number is 29. Find the number.

66. Twice a number, less seven, is 17. Find the number.

67. A number divided by five, less six, is four. Find the number.

68. Sue has three times as much money as Jane. Together they have 64 dollars. How much does each have?

69. Al is seven years older than Maria. The sum of their ages is 51. What is each of their ages?

70. Six times Glen's age plus two equals four times his age plus 20. Find his age.

55.	
56.	
57.	
58.	
59.	
60.	
61.	
62.	
63.	
64.	
65.	
66.	
67.	
68.	
69.	
70.	

70

71.	
72.	
73.	
74.	
75.	
76.	
77.	
78.	
79.	
80.	
81.	
82.	
83.	
84.	
85.	
86.	

Use the following information to answer 71 - 74.

There are 6 green marbles, 5 red marbles, 4 white marbles, and 1 blue marble in a can. What is the probability for each of the following?

71. a red marble

72. a green marble

73. a green or blue marble

74. not a red marble

Use the spinner to find the probability of spinning once and landing on each of the following.

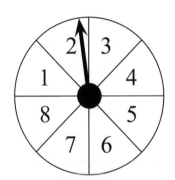

75. a seven.

76. an even number.

77. a number greater than four.

78. a one, a three, or a five.

Arrange the data in order from least to greatest, then answer the questions.

2, 7, 3, 10, 3

79. What is the range?

80. What is the mode?

81. What is the mean?

82. What is the median?

4, 10, 4, 2, 6, 14, 16

83. What is the mode?

84. What is the mean?

85. What is the range?

86. What is the median?

Final Review - All Pre-Algebra Concepts

Solve each of the following problems.

87. $\dfrac{7}{8}$

$+\dfrac{3}{8}$

88. $7\dfrac{1}{4}$

$-3\dfrac{3}{4}$

89. $3\dfrac{3}{5}$

$+2\dfrac{1}{10}$

90. $\dfrac{5}{8} \times 3\dfrac{1}{5} =$

91. $2\dfrac{1}{2} \times 3\dfrac{1}{2} =$

92. $2\dfrac{1}{3} \div \dfrac{1}{2} =$

93. $5\dfrac{1}{2} \div 1\dfrac{1}{2} =$

94. .6 + 7.62 + 5.2 + 6 =

95. 6.3 - 1.275 =

96. 72 - 1.68 =

97. 2.19
$\times 7$

98. .36
$\times 1.2$

99. $5\overline{)6.7}$

100. $.15\overline{).0045}$

87.	
88.	
89.	
90.	
91.	
92.	
93.	
94.	
95.	
96.	
97.	
98.	
99.	
100.	

72

Solutions

Page 4

Review Exercises
1. 1,159
2. 436
3. 2,282

S1. No, members are countable
S2. No, members are uncountable
1. No, 3 is a member of each
2. Yes, can be paired 1-1
3. Answers vary
4. Answers vary
5. {3,5,7,9,11}
6. {0,2,4,6,8,10,12}
7. {3,4,5,6,7,8,9}
8. {10,15,20,25,30}
9. {1,3,5}
10. {8,9,10,11,12}

Problem Solving: 9 girls

Page 5

Review Exercises
1. Answers vary
2. Answers vary
3. Answers vary
4. Answers vary
5. No, 10 is a member of both sides
6. No, cannot be paired in a 1-1 correspondence

S1. Yes, all members of A are members of B
S2. {5,6,7}
1. {1,2,3,4,5,6,7,8}
2. No, not all members of A are members of B
3. {5}, {6}, {7}, {5,6}, {5,7}, {6,7}, {5,6,7}
4. {1,2,4,5,7}
5. {1,2,4,8}
6. {1,2,3,4,5,6,7}
7. {1,2,3,4,5,6,7,8,10}
8. {1,2,4,6}
9. Yes, can be paired in a 1-1 correspondence
10. No, 6 is common to both sets

Problem Solving: 80 cards

Page 6

Review Exercises
1. {2}
2. {1,2,3,4,6,8}
3. {1,2,3,6}
4. {2}
5. No, cannot be paired in a 1-1 correspondence
6. No, the members are countable

S1. 3
S2. -21
1. 14
2. -18
3. -14
4. -31
5. 24
6. 37
7. -168
8. -13
9. -34
10. -10

Problem Solving: -20°

Page 7

Review Exercises
1. -7
2. 13
3. -39
4. 0
5. A set is a well defined collection of objects
6. A set whose number of members is countable.

S1. -4
S2. -7
1. -2
2. -7
3. -8
4. 14
5. -2
6. 10
7. -22
8. -13
9. -25
10. -84

Problem Solving: $55

Page 8

Review Exercises
1. Ø
2. {0,1,4,5,8,9,10,12,15}
3. {10,15}
4. {1,4,8,9,10,11,12,15}
5. {1,4,8,9,12}
6. Ø

S1. -14
S2. -3
1. 12
2. -3
3. 9
4. -28
5. 46
6. -48
7. -10
8. -11
9. -103
10. 15

Problem Solving: 27 ft.

Page 9

Review Exercises
1. -56
2. 22
3. -35
4. -9
5. -11
6. 4

S1. 48
S2. -126
1. 68
2. -64
3. 288
4. -368
5. -736
6. -24
7. -32
8. 72
9. 72
10. 330

Problem Solving: Floor 31

Page 10

Review Exercises
1. -11
2. -56
3. -2
4. 2
5. -35
6. 36

S1. -4
S2. 6
1. -16
2. 48
3. 15
4. -26
5. -3
6. -2
7. 1
8. -1
9. 3
10. 2

Problem Solving: -11°

Page 11

Review Exercises
1. -2
2. 2
3. -16
4. -1
5. -19
6. -2
7. 13
8. -12
9. -27
10. -1
11. -48
12. 76
13. 56
14. 48
15. -9
16. 42
17. 16
18. 3
19. -2
20. -20

Solutions

Page 12	Page 13	Page 14	Page 15
Review Exercises	Review Exercises	Review Exercises	Review Exercises
1. {1,4}	1. -59	1. .4	1. 49
2. {1,2,3,4,6,7}	2. -36	2. -5.9	2. 729
3. {1,6}	3. -64	3. -7.8	3. 36
4. {1,2,3,4,5,6,9}	4. -28	4. -5/6	4. 4
5. {1,3,4,5,6,7,9}	5. 4	5. -1/10	5. 1
6. {1,3,4,5,9}	6. 28	6. 3/5	6. 9
S1. 3/10	S1. -.91	S1. 16	S1. 12^3
S2. -9/10	S2. -10.3	S2. -27	S2. 3^3
1. -1/4	1. -12.81	1. 216	1. 2^6
2. -1 1/6	2. 3.17	2. 1	2. $(-9)^3$
3. -2	3. -4.284	3. 16	3. 16^4
4. 3/8	4. 12.13	4. 32	4. 7^2 or $(-7)^2$
5. -7/12	5. 2.37	5. 7	5. 10^2 or $(-10)^2$
6. 2 1/2	6. .18	6. 512	6. 11^2 or $(-11)^2$
7. 1 1/12	7. .426	7. -1	7. $(-1)^4$
8. 1/8	8. -6.93	8. 3,125	8. 2^5
9. -10	9. 2.01	9. -125	9. 2^4 or 4^2 or $(-2)^4$ or $(-4)^2$
10. 7/15	10. -17.04	10. 81	10. 9^6
Problem Solving: 50 sixth graders	Problem Solving: 118 pounds	Problem Solving: 2	Problem Solving: (-3)

Page 16	Page 17	Page 18	Page 19
Review Exercises	Review Exercises	Review Exercises	Review Exercises
1. -9	1. 36	1. 13^4	1. 49
2. -3	2. -8	2. 2^7	2. 6
3. -7/12	3. 6^4	3. 2^6 or 8^2 or $(-2)^6$ or $(-8)^2$	3. -2
4. 3.6	4. 8	4. $(-2)^4$	4. -56
5. -1.04	5. 13	5. 2^3	5. -4
6. -3/8	6. 11	6. 10^2 or $(-10)^2$	6. 24
		7. 4	
S1. 5	S1. 36	8. 8	S1. 28
S2. 12	S2. 8	9. 5	S2. 38
1. 4	1. 2	10. 20	1. 7
2. 11	2. 1	11. 3	2. 15
3. 1	3. 25	12. 18	3. 34
4. 30	4. 88	13. 22	4. 25
5. 10	5. 144	14. 4	5. 12
6. 20	6. 64	15. 85	6. 27
7. 13	7. 54	16. 400	7. 14
8. 3	8. 4	17. 63	8. 4
9. 16	9. 144	18. 675	9. 28
10. 40	10. 9	19. 25	10. 36
		20. 10	
Problem Solving: -41	Problem Solving: 81		Problem Solving: -1 yard

Solutions

Page 20

Review Exercises
1. {2,4}
2. {8}
3. Ø
4. {1,2,4,5,7,8,9}
5. {1,2,4,5,6,8,10}
6. {1,2,4,5,8}

S1. 4
S2. 9
1. 29
2. 6
3. 2
4. 32
5. 1
6. 7
7. 6
8. 12
9. 7
10. 24

Problem Solving: $29

Page 21

Review Exercises
1. 27
2. 28
3. -13
4. A well-defined collection of objects
5. -1 1/4
6. -.41

S1. commutative (addition)
S2. distributive
1. inverse property (addition)
2. associative (multiplication)
3. identity (addition)
4. inverse (multiplication)
5. commutative (addition)
6. commutative (multiplication)
7. associative (addition)
8. identity (multiplication)
9. distributive
10. inverse (addition)

Problem Solving: 19

Page 22

Review Exercises
1. 11
2. 17
3. 10
4. 23
5. 60
6. 42

S1. $3 + (7 + 9)$
S2. 15×7
1. 1
2. $3 \times 6 + 3 \times 2$
3. $12 + 9$
4. $(3 \times 9) \times 5$
5. $3(5 + 7)$
6. -9
7. 1
8. 5
9. $(3 + 5) + 6$
10. $3 \times 4 + 3 \times (-2)$

Problem Solving: $42

Page 23

Review Exercises
1. 10
2. 125
3. 21
4. -90
5. 5
6. -5/6

S1. 2.3×10^9
S2. 1.49×10^{-7}
1. 6.53×10^{11}
2. 1.597×10^5
3. 1.06×10^8
4. 7.216×10^{-6}
5. 1.096×10^9
6. 1.963×10^{-3}
7. 1.6×10^{-10}
8. 8×10^{-10}
9. 7×10^{12}
10. 1.287×10^{-7}

Problem Solving:
1.86×10^5 miles per second

Page 24

Review Exercises
1. 1.23×10^5
2. 3.21×10^{-4}
3. distributive
4. -17
5. -3
6. 35

S1. 7,032,000
S2. .000056
1. 230,000
2. .0000000913
3. .000012362
4. 517,000,000,000
5. 1,127
6. .003012
7. 6,670,000
8. 21,000
9. .00000007
10. 8,000,000

Problem Solving:
93,000,000 miles

Page 25

Review Exercises
1. 1.23×10^5
2. 5.6×10^{-6}
3. 2,760,000
4. .0000375
5. answers vary
6. answers vary

S1. 5/3
S2. 9/2
1. 7/2
2. 6/5
3. 6/5
4. 5/1
5. 6/5
6. 4/3
7. 7/3
8. 3/2
9. 1/2
10. 3/1

Problem Solving: 12/5

Page 26

Review Exercises
1. 2.7×10^{-4}
2. 2.916×10^6
3. 721,000
4. .0000623
5. 30
6. -1.32

S1. yes
S2. no
1. yes
2. no
3. yes
4. yes
5. yes
6. no
7. yes
8. no
9. yes
10. no

Problem Solving: 2

Page 27

Review Exercises
1. 5/3
2. yes $(4 \times 10 = 8 \times 5)$
3. no $(5 \times 5 \neq 3 \times 7)$
4. 16
5. 48
6. 275

S1. 1
S2. 16
1. 3
2. 2
3. 42
4. 6 2/5
5. 6
6. 21
7. 18
8. 2 4/5
9. 1.2
10. 5

Problem Solving: -32°

Solutions

Page 28

Review Exercises
1. yes ($4 \times 9 = 3 \times 12$)
2. n = 30
3. n = 12
4. 2.34×10^8
5. $2.35 \times 10\text{-}3$
6. 720,000

S1. 204 miles
S2. $12
1. 2 gallons
2. $17.50
3. 15 girls
4. 40 miles
5. 2 4/5 pounds

Problem Solving: 12°

Page 29

Review Exercises
1. 3/1
2. 12/5
3. 8/3
4. yes, $15 \times 24 = 12 \times 30$
5. no, $7 \times 9 \neq 8 \times 8$
6. yes, $5 \times 9 = 3 \times 15$
7. 4
8. 33
9. 5
10. 35
11. 4 1/2
12. 9
13. 20
14. 3
15. 9
16. $8.40
17. 30 boys
18. 70 miles
19. 2 gallons
20. 4 pounds

Page 30

Review Exercises
1. 21
2. 3 3/5
3. 9
4. 18
5. 26
6. -33

S1. .2, 1/5
S2. .09, 9/100
1. .16, 4/25
2. .06, 3/50
3. .75, 3/4
4. .4, 2/5
5. .01, 1/100
6. .45, 9/20
7. .12, 3/25
8. .05, 1/20
9. .5, 1/2
10. .13, 13/100

Problem Solving: 19/20

Page 31

Review Exercises
1. .8
2. .07
3. 1/4
4. 109.2
5. 128
6. 18

S1. 17.5
S2. 150
1. 4.32
2. 51
3. 15
4. 112.5
5. 32
6. 80
7. 10
8. 216
9. 112.5
10. 13.2

Problem Solving: 34 correct

Page 32

Review Exercises
1. 46.5
2. 24
3. .75
4. 70%
5. 135
6. 600

S1. 25%
S2. 75%
1. 25%
2. 80%
3. 50%
4. 90%
5. 60%
6. 75%
7. 75%
8. 75%
9. 80%
10. 95%

Problem Solving: 75%

Page 33

Review Exercises
1. 3.2
2. 32
3. 75%
4. 90%
5. 50.04
6. 200

S1. 20
S2. 30
1. 48
2. 80
3. 25
4. 4
5. 15
6. 20
7. 60
8. 75
9. 45
10. 125

Problem Solving: 25 students

Page 34

Review Exercises
1. .0072
2. 2 19/10,000
3. 60%
4. 15
5. .021
6. 80

S1. 20 questions
S2. 75%
1. $25
2. $1,600
3. 30
4. 90%
5. 250 cows
6. 25
7. $240
8. 180 boys
9. 20%
10. $210

Problem Solving: $57,600

Page 35

Review Exercises
1. 32.744
2. 2.358
3. 21.98
4. .01248
5. .48
6. 8.1

S1. 30
S2. 30
1. 20%
2. 90%
3. 30 students
4. 18 passes
5. 20% are red
6. 75%
7. 250 students
8. $25
9. 60%
10. $691.20

Problem Solving: 97

Solutions

Page 36

Review Exercises
1. 13%
2. 3%
3. 70%
4. 19%
5. 60%
6. .08, 2/25
7. .18, 9/50
8. .8, 4/5
9. 2.22
10. 128
11. 80%
12. 75%
13. 12
14. 75
15. 80%
16. 60%
17. 256 girls
18. 26 games
19. 75%
20. 150 students

Page 37

Review Exercises
1. 70%
2. 80%
3. 3/25
4. 12
5. 25%
6. 25

S1. 1, 30, 2, 15, 3, 10, 5, 6
S2. 1, 36, 2, 18, 3, 12, 4, 9, 6
1. 1, 100, 2, 50, 4, 25, 5, 20, 10
2. 1, 42, 2, 21, 3, 14, 6,7
3. 1, 70, 2, 35, 5, 14, 7, 10
4. 1, 81, 3, 27, 9
5. 1, 50, 2, 25, 5, 10
6. 1, 40, 2, 20, 4, 10, 5, 8
7. 1, 75, 3, 25, 5, 15
8. 1, 90, 2, 45, 3, 30, 5, 18, 6, 15, 9, 10
9. 1, 20, 2, 10, 4, 5
10. 1, 50, 2, 25, 5, 10

Problem Solving: 54 correct

Page 38

Review Exercises
1. -18
2. 24
3. 20
4. 7 1/2
5. 15
6. 25%

S1. 2
S2. 4
1. 2
2. 3
3. 14
4. 16
5. 20
6. 10
7. 5
8. 12
9. 12
10. 20

Problem Solving:
5,879,000,000,000 miles

Page 39

Review Exercises
1. 1.2×10^{-6}
2. 4.96×10^8
3. 13,200,000
4. .00000464
5. 1,60,2,30,3,10,5,6
6. 4

S1. 4, 6, 8, 10
S2. 0, 12, 18, 30
1. 10, 15, 20, 25
2. 0, 6, 12, 15
3. 0, 30, 40, 50
4. 0, 4, 8
5. 22, 44
6. 24, 32, 40
7. 60, 80, 100
8. 14, 28, 35
9. 90, 120, 150
10. 27, 45

Problem Solving: 120 pitches

Page 40

Review Exercises
1. 1, 30, 2, 15, 3, 10, 5, 6
2. 4
3. 0, 8, 16, 24, 32, 40
4. 3
5. 25%
6. 35

S1. 12
S2. 24
1. 15
2. 30
3. 60
4. 30
5. 36
6. 60
7. 48
8. 40
9. 36
10. 60

Problem Solving: $12.96

Page 41

1. 1, 24, 2, 12, 3, 8, 4, 6
2. 1, 16, 2, 8, 4
3. 1, 32, 2, 16, 4, 8
4. 1, 28, 2, 14, 4, 7
5. 1, 70, 2, 35, 7, 10
6. 1, 25, 5
7. 4
8. 12
9. 20
10. 5
11. 7
12. 18
13. 9, 12, 15
14. 9, 27, 36, 45
15. 15, 30, 45, 60
16. 12
17. 60
18. 60
19. 12
20. 24

Page 42

Review Exercises
1. 8
2. 12
3. 1, 28, 2, 14, 4, 7
4. 0, 12, 24, 36, 48, 60
5. yes, $3 \times 8 = 4 \times 6$
6. 9

S1. 0
S2. -7, -2, 10
1. 5, 9
2. 3, 4
3. 2, 4, 7
4. -5, -8
5. 10, 9, -6, 6
6. 9, -7, 1
7. -8, 8, 0, -3
8. 0, 7, 8
9. -6, 4, -3
10. 5, -3, 9, -8

Problem Solving: 8°

Page 43

Review Exercises
1. {4,6,8,10}
2. {1,3,4,5,6,8,9,10}
3. {2,4,5,6,8,9,10}
4. {4,5,6,8,10}
5. no, cannot be paired in 1-1 correspondence
6. no, they have members in common

S1. 1
S2. 7
1. 3
2. 5
3. 5
4. 18
5. 4
6. 11
7. 8
8. 7
9. 4
10. 12

Problem Solving: 512 miles

Page 44	Page 45	Page 46	Page 47

Page 44

Review Exercises
1. 7
2. -22
3. 8
4. -42
5. 5
6. -2

S1. (2, 1)
S2. (-4, 2)
1. (6, 3)
2. (2, -5)
3. (-7, -3)
4. (-5, 1)
5. (4, 6)
6. (4, -3)
7. (-3, -5)
8. (-2, -2)
9. (2, 1)
10. (-6, 7)

Problem Solving: $24

Page 45

Review Exercises
1. -7/15
2. -.68
3. 3/8
4. 1
5. -5
6. 2

S1. B
S2. A
1. C
2. D
3. F
4. M
5. E
6. J
7. H
8. G
9. K
10. I

Problem Solving: 95%

Page 46

Review Exercises
1. 32
2. 22
3. 5
4. 72
5. 1.7×10^{-4}
6. 2.13×10^{5}

S1. 1/3
S2. 3/2
1. -3/2
2. 1/3
3. -2
4. 5/8
5. 1/2
6. 7/2
7. 6/5
8. 4/5
9. 3/2
10. -3/2

Problem Solving: 320 girls

Page 47

1. -4
2. 9, 0, 10
3. 3, 2, -7
4. 8, 1, -4, -8
5. 3
6. 6
7. 5
8. 8
9. (5, 3)
10. (-6, 1)
11. (4, -4)
12. (-6, -5)
13. (2, 6)
14. 5/3
15. B
16. I
17. H
18. D
19. A
20. 1/8

Page 48

Review Exercises
1. 6
2. -18
3. -24
4. -16
5. -7
6. -8

Problem Solving: $6.00

S1.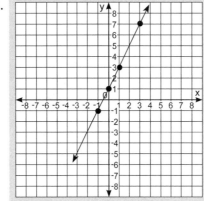

S2.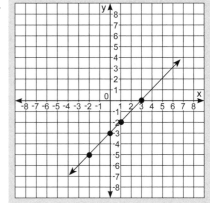

1.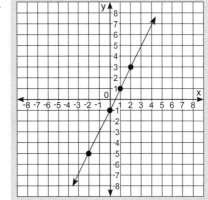

2.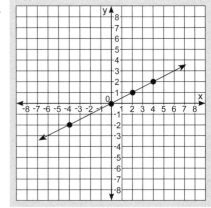

Solutions

Page 49

Review Exercises
1. 8 2/5
2. 3
3. 75
4. -1/2
5. -3
6. -2.1

Problem Solving: 80%

S1.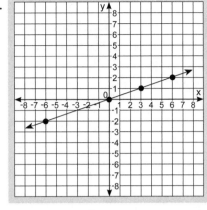

S2.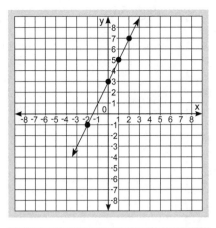

1.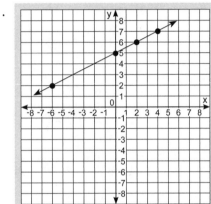

2.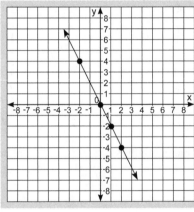

Page 50	Page 51	Page 52	Page 53
Review Exercises	Review Exercises	Review Exercises	Review Exercises
1. 1.72×10^6	1. 10	1. 1	1. 1/10
2. 3.8×10^{-7}	2. 23	2. 72	2. -4/5
3. 196,300,000	3. 102	3. 2	3. -1 1/3
4. .00034	4. -8	4. 2.1×10^5	4. -3
5. -10	5. 6	5. 3.16×10^{-3}	5. -.64
6. .19	6. 21	6. 16	6. -1.1
S1. 5	S1. 12	S1. 7	S1. 3
S2. 32	S2. 3	S2. 4	S2. 45
1. -3	1. -14	1. -1	1. 8
2. 40	2. -12	2. -4	2. -3
3. -13	3. -9	3. -15	3. 10
4. -5	4. -5	4. 75	4. -3
5. 13	5. 18	5. 12	5. 5
6. -12	6. -6	6. 1	6. -4
7. 34	7. 2	7. 2	7. -10
8. 4	8. -20	8. 50	8. -7
9. -11	9. -15	9. 10	9. -3
10. -27	10. 14	10. -3	10. 10
Problem Solving: 7/3	Problem Solving: 9/4	Problem Solving: 200 students	Problem Solving: 14

Page 54

Review Exercises
1. 216
2. 1
3. 9
4. 13
5. 12
6. 158

S1. -2
S2. 3
1. 7
2. 2
3. 4
4. 4
5. 3
6. 1
7. -1
8. 4
9. 2
10. 3

Problem Solving: 20%

Page 55

Review Exercises
1. 20
2. no, $4 \times 5 \neq 3 \times 7$
3. 8/3
4. 60
5. 25%
6. 35

S1. 4
S2. 5
1. -2
2. -3
3. 10
4. -5
5. 12
6. 1
7. 7
8. 3
9. -25
10. -10

Problem Solving:
8 blue marbles

Page 56

Review Exercises
1. 1, 48, 2, 24, 3, 16,
   4, 12, 6, 8
2. 8
3. 30
4. 5
5. 20
6. -5

S1. $2x - 7 = 12$
S2. $3x + 2 = 30$
1. $2x + 5 = 14$
2. $4x - 6 = 10$
3. $4x - 5 = 12$
4. $x/3 - 4 = 2x + 8$
5. $2(x + 2) = 10$
6. $5x - 3 = 17$
7. $2x - 6 = 15$
8. $3x - 2 = 2x + 7$
9. $x + 4 = 7 + -12$
10. $n/5 = 25$

Problem Solving: 227.5 miles

Page 57

Review Exercises
1. 7
2. -4
3. 12
4. 5
5. 15
6. 3

S1. 11
S2. 12
1. 36
2. -3
3. 7
4. 4
5. 3

Problem Solving: $12,500

Page 58

Review Exercises
1. .000000361
2. 1.27×10^{-6}
3. 7.29×10^{8}

S1. Kevin is 12
    Amir is 18
S2. Short piece = 11 inches
    Long piece = 33 inches
1. Bob earned $20
   Bill earned $46
2. Monday, $90
   Tuesday, $120
3. 12 years old
4. Weekly salary is $140
5. John is 30

Problem Solving: 89

Page 59

Review Exercises
1. -11
2. 20
3. 44
4. -9
5. 3
6. 11

S1. 6
S2. Bert is 110 pounds
    Bob is 160 pounds
1. -9
2. 6
3. -24
4. Ellen is 11
   Roy is 33
5. 3

Problem Solving:
600 miles per hour

Page 60

1. -7
2. 13
3. 56
4. 3
5. -5
6. -18
7. -7
8. 15
9. 8
10. 8
11. -11
12. 21
13. 12
14. -5
15. 6
16. 6
17. Sue, $22
    Ann, $44
18. Ron is 28
    Bill is 36
19. -2
20. 18

Page 61

Review Exercises
1. 0,8,16,24,32,40,48
2. 1, 60, 2, 30, 3, 20, 4, 15,
   5, 12, 6, 10
3. 20
4. 6×10^{-6}
5. 2.1×10^{6}
6. .0021

S1. 3/12 = 1-4
S2. 7/12
1. 2/12 = 1/6
2. 1/12
3. 9/12 = 3/4
4. 10/12 = 5/6
5. 9/12 = 3/4
6. 5/12
7. 10/12 = 5/6
8. 6/12 = 1/2
9. 6/12 = 1/2
10. 4/12 = 1/3

Problem Solving: -3

Solutions

Page 62

Review Exercises
1. -10
2. -25
3. -10
4. 4
5. 15
6. 9

S1. 1/8
S2. 4/8 = 1/2
1. 1/8
2. 7/8
3. 4/8 = 1/2
4. 4/8 = 1/2
5. 2/8 = 1/4
6. 0/8
7. 2/8 = 1/4
8. 5/8
9. 5/8
10. 4/8 = 1/2

Problem Solving: 20 pounds

Page 63

Review Exercises
1. 30%
2. 3%
3. 60%
4. 2
5. 25%
6. 20

S1. range 7, mode 4
S2. range 6, mode 6
1. range 6, mode 7
2. range 23, mode 30
3. range 10, mode 3
4. range 19, mode 9
5. range 7, mode 3
6. range 11, mode 8
7. range 9, mode 2
8. range 6, mode 9
9. range 9, mode 2
10. range 19, mode 2

Problem Solving: $.38

Page 64

Review Exercises
1. 8/5
2. no, $7 \times 11 \neq 8 \times 9$
3. 20
4. 1.28×10^6
5. 9.62×10^{-5}
6. .000062

S1. mean 3, median 3
S2. mean 4, media 2
1. mean 3, median 2
2. mean 3, median 2
3. mean 15, median 15
4. mean 2, median 1
5. mean 6, median 6
6. mean 124, median 126
7. mean 4, median 4
8. mean 4, median 4
9. mean 5, median 4
10. mean 50, median 50

Problem Solving: 32 students

Page 65

Review Exercises
1. 21
2. 14
3. 15
4. 25
5. 20
6. 11

S1. 6
S2. 2
1. 5
2. 6
3. 3
4. 2
5. 8
6. 4
7. 10
8. 2
9. 5
10. 5

Problem Solving:
186,000 miles per second

Page 66

1. 3/10
2. 4/10 = 2/5
3. 5/10 = 1/2
4. 7/10
5. 8/10 = 4/5
6. 6/10 = 3/5
7. 1/8
8. 4/8 = 1/2
9. 5/8
10. 2/8 = 1/4
11. 4/8 = 1/2
12. 2/8 = 1/4
13. 8
14. 4
15. 5
16. 4
17. 2
18. 4
19. 7
20. 3

Page 67

1. {2,3,4}
2. {0,1,2,3,4,5,6,8,9}
3. {1,2,4,5}
4. 3
5. -23
6. 36
7. 8
8. -.55
9. -9/10
10. 125
11. 7
12. 33
13. 22
14. 68
15. 2
16. 20
17. commutative (addition)
18. distributive
19. 1.28×10^9
20. 6.53×10^{-6}

Page 68

21. 60,900,000
22. .00000762
23. 9/5
24. no, $8 \times 3 \neq 4 \times 7$
25. 21
26. 6 blue marbles
27. 4
28. 25%
29. 32
30. 24
31. 40%
32. 30
33. 1, 40, 2, 20, 4, 10, 5, 8
34. 20
35. 0, 6, 12, 18, 24, 30, 36
36. 24
37. -8, -2, 0
38. -5, 9, -1, -7
39. -6, 4, -3
40. 9, 8, 1

Page 69

41. (6, -3)
42. (-5, 1)
43. (4, -3)
44. (-4, -2)
45. (4, 6)
46. H
47. G
48. L
49. J
50. K
51. 2/3
52. 3/8

53.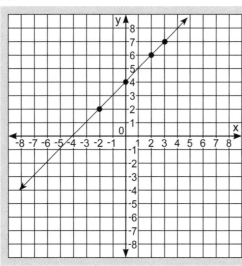

54.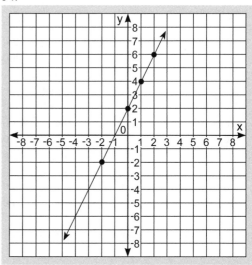

Page 70	Page 71	Page 72
55. 9	71. 5/6	87. 1/2
56. -15	72. 6/16 = 3/8	88. 3 1/2
57. 18	73. 7/16	89. 6 3/10
58. -3	74. 11/16	90. 2
59. 6	75. 1/8	91. 8 3/4
60. -3	76. 4/8 = 1/2	92. 12 2/3
61. 24	77. 4/8 = 1/2	93. 3 2/3
62. 4	78. 3/8	94. 19.42
63. -6	79. 8	95. 5.025
64. 1	80. 3	96. 70.32
65. 9	81. 5	97. 15.33
66. 12	82. 3	98. .432
67. 50	83. 4	99. 1.34
68. Jane, $16	84. 8	100. .03
Sue, $48	85. 14	
69. Maria is 22	86. 6	
Al is 29		
70. 9		

Glossary

A

absolute value   The distance of a number from 0 on the number line. The absolute value is always positive.

acute angle   An angle with a measure of less than 90 degrees.

adjacent   Next to.

algebraic expression   A mathematical expression that contains at least one variable.

angle   Any two rays that share an endpoint will form an angle.

associative properties   For any a, b, c:
addition: $(a + b) + c = a + (b + c)$
multiplication: $(ab)c = a(bc)$

B

base   The number being multiplied. In an expression such as 4^2, 4 is the base.

C

coefficient   A number that multiplies the variable. In the term 7x, 7 is the coefficient of x.

commutative properties   For any a, b:
addition: $a + b = b + a$
multiplication: $ab = ba$

complementary angles   Two angles that have measures whose sum is 90 degrees.

congruent   Two figures having exactly the same size and shape.

coordinate plane   The plane which contains the x- and y-axes. It is divided into 4 quadrants. Also called coordinate system and coordinate grid.

coordinates   An ordered pair of numbers that identify a point on a coordinate plane.

D

data   Information that is organized for analysis.

degree   A unit that is used in measuring angles.

denominator   The bottom number of a fraction that tells the number of equal parts into which a whole is divided.

disjoint sets   Sets that have no members in common. {1,2,3} and {4,5,6} are disjoint sets.

Glossary

distributive property   For real numbers a, b, and c: $a(b + c) = ab + ac$.

E

element of a set   Member of a set.

empty set   The set that has no members. Also called the null set and written $\emptyset$ or { }.

equation   A mathematical sentence that contains an equal sign (=) and states that one expression is equal to another expression.

equivalent   Having the same value.

exponent   A number that indicates the number of times a given base is used as a factor. In the expression n^2, 2 is the exponent.

expression   Variables, numbers, and symbols that show a mathematical relationship.

extremes of a proportion   In the proportion $\dfrac{a}{b} = \dfrac{c}{d}$, a and d are the extremes.

F

factor   An integer that divides evenly into another.

finite   Something that is countable.

formula   A general mathematical statement or rule. Used often in algebra and geometry.

function   A set of ordered pairs that pairs each x-value with one and only one y-value. (0,2), (-1,6), (4,-2), (-3,4) is a function.

G

graph   To show points named by numbers or ordered pairs on a number line or coordinate plane. Also, a drawing to show the relationship between sets of data.

greatest common factor   The largest common factor of two or more numbers. Also written GCF. The greatest common factor of 15 and 25 is 5.

grouping symbols   Symbols that indicate the order in which mathematical operations should take place. Examples include parentheses ( ), brackets [ ], braces { }, and fraction bars —— .

H

hypotenuse   The side opposite the right angle in a right triangle.

I

identity properties of addition and multiplication   For any real number a:
addition: $a + 0 = 0 + a = a$
multiplication: $1 \times a = a \times 1 = a$

inequality   A mathematical sentence that states one expression is greater than or less than another. Inequality symbols are read as follows: $<$ less than
$\leq$ less than or equal to
$>$ greater than
$\geq$ greater than or equal to

infinite   Having no boundaries or limits. Uncountable.

integers   Numbers in a set. ...-3, -2, -1, 0, 1, 2, 3...

intersection of sets   If A and B are sets, then A intersection B is the set whose members are included in both sets A and B, and is written $A \cap B$. If set A = {1,2,3,4} and set B = {1,3,5}, then $A \cap B$ = {1,3}

inverse properties of addition and multiplication   For any number a:
addition: $a + -a = 0$
multiplication: $a \times 1/a = 1$ $(a \neq 0)$

inverse operations   Operations that "undo" each other. Addition and subtraction are inverse operations, and multiplication and division are inverse operations.

L

least common multiple   The least common multiple of two or more whole numbers is the smallest whole number, other than zero, that they all divide into evenly. Also written as LCM. The least common multiple of 12 and 8 is 24.

linear equation   An equation whose graph is a straight line.

M

mean   In statistics, the sum of a set of numbers divided by the number of elements in the set. Sometimes referred to as average.

means of a proportion   In the proportion $\frac{a}{b} = \frac{c}{d}$, b and c are the means.

median   In statistics, the middle number of a set of numbers when the numbers are arranged in order of least to greatest. If there are two middle numbers, find their mean.

mode   In statistics, the number that appears most frequently. Sometimes there is no mode. There may also be more than one mode.

multiple   The product of a whole number and another whole number.

Glossary

natural numbers   Numbers in the set 1, 2, 3, 4,... Also called counting numbers.

negative numbers   Numbers that are less than zero.

null set   The set that has no members. Also called the empty set and written Ø or { }.

number line   A line that represents numbers as points.

numerator   The top part of a fraction.

O

obtuse angle   An angle whose measure is greater than 90° and less than 180°.

opposites   Numbers that are the same distance from zero, but are on opposite sides of zero on a number line. 4 and -4 are opposites.

order of operations   The order of steps to be used when simplifying expressions.
1. Evaluate within grouping symbols.
2. Eliminate all exponents.
3. Multiply and divide in order from left to right.
4. Add and subtract in order from left to right.

ordered pair   A pair of numbers (x,y) that represent a point on the coordinate plane. The first number is the x-coordinate and the second number is the y-coordinate.

origin   The point where the x-axis and the y-axis intersect in a coordinate plane. Written as (0,0).

outcome   One of the possible events in a probability situation.

P

parallel lines   Lines in a plane that do not intersect. They stay the same distance apart.

percent   Hundredths or per hundred. Written %.

perimeter   The distance around a figure.

perpendicular lines   Lines in the same plane that intersect at a right (90°) angle.

pi   The ratio of the circumference of a circle to its diameter. Written π. The approximate value for π is 3.14 as a decimal and $\frac{22}{7}$ as a fraction.

plane   A flat surface that extends infinitely in all directions.

point   An exact position in space. Points also represent numbers on a number line or coordinate plane.

positive number   Any number that is greater than 0.

power   An exponent.

prime number   A whole number greater than 1 whose only factors are 1 and itself.

probability   What chance, or how likely it is for an event to occur. It is the ratio of the ways a certain outcome can occur and the number of possible outcomes.

proportion   An equation that states that two ratios are equal. $\frac{4}{8} = \frac{2}{4}$ is a proportion.

Pythagorean theorem   In a right triangle, if c is the hypotenuse, and a and b are the other two legs, then $a^2 + b^2 = c^2$.

Q ▓▓

quadrant   One of the four regions into which the x-axis and y-axis divide a coordinate plane.

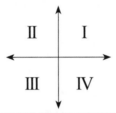

R ▓▓

range   The difference between the greatest number and the least number in a set of numbers.

ratio   A comparison of two numbers using division. Written a:b, a to b, and a/b.

reciprocals   Two numbers whose product is 1. $\frac{2}{3}$ and $\frac{3}{2}$ are reciprocals because $\frac{2}{3} \times \frac{3}{2} = 1$.

reduce   To express a fraction in its lowest terms.

relation   Any set of ordered pairs.

right angle   An angle that has a measure of 90°.

rise   The change in y going from one point to another on a coordinate plane. The vertical change.

run   The change in x going from one point to another on a coordinate plane. The horizontal change.

S ▓▓

scientific notation   A number written as the product of a numbers between 1 and 10 and a power of ten. In scientific notation, $7,000 = 7 \times 10^3$.

set   A well-defined collection of objects.

slope   Refers to the slant of a line. It is the ratio of rise to run.

Glossary

solution   A number that can be substituted for a variable to make an equation true.

square root   Written $\sqrt{}$. The $\sqrt{36} = 6$ because $6 \times 6 = 36$.

statistics   Involves data that is gathered about people or things and is used for analysis.

subset   If all the members of set A are members of set B, then set A is a subset of set B. Written $A \subset B$. If set A = {1,2,3} and set B = {0,1,2,3,5,8}, set A is a subset of set B because all of the members of a set A are also members of set B.

U

union of sets   If A and B are sets, the union of set A and set B is the set whose members are included in set A, or set B, or both set A and set B. A union B is written $A \cup B$. If set = {1,2,3,4} and set B = {1,3,5,7}, then $A \cup B$ = {1,2,3,4,5,7}.

universal set   The set which contains all the other sets which are under consideration.

V

variable   A letter that represents a number.

Venn diagram   A type of diagram that shows how certain sets are related.

vertex   The point at which two lines, line segments, or rays meet to form an angle.

W

whole number   Any number in the set 0, 1, 2, 3, 4...

X

x-axis   The horizontal axis on a coordinate plane.

x-coordinate   The first number in an ordered pair. Also called the abscissa.

Y

y-axis   The vertical axis on a coordinate plane.

y-coordinate   The second number in an ordered pair. Also called the ordinate.

Important Symbols

<	less than		π	pi
≤	less than or equal to		{ }	set
>	greater than		| |	absolute value
≥	greater than or equal to		$.\overline{n}$	repeating decimal symbol
=	equal to		1/a	the reciprocal of a number
≠	not equal to		%	percent
≅	congruent to		(x,y)	ordered pair
( )	parenthesis		⊥	perpendicular
[ ]	brackets		| |	parallel to
{ }	braces		∠	angle
...	and so on		∈	element of
• or ×	multiply		∉	not an element of
∞	infinity		∩	intersection
a^n	the nth power of a number		∪	union
√	square root		⊂	subset of
Ø, { }	the empty set or null set		⊄	not a subset of
∴	therefore		△	triangle
°	degree			

Multiplication Table

x	2	3	4	5	6	7	8	9	10	11	12
2	4	6	8	10	12	14	16	18	20	22	24
3	6	9	12	15	18	21	24	27	30	33	36
4	8	12	16	20	24	28	32	36	40	44	48
5	10	15	20	25	30	35	40	45	50	55	60
6	12	18	24	30	36	42	48	54	60	66	72
7	14	21	28	35	42	49	56	63	70	77	84
8	16	24	32	40	48	56	64	72	80	88	96
9	18	27	36	45	54	63	72	81	90	99	108
10	20	30	40	50	60	70	80	90	100	110	120
11	22	33	44	55	66	77	88	99	110	121	132
12	24	36	48	60	72	84	96	108	120	132	144

Commonly Used Prime Numbers

2	3	5	7	11	13	17	19	23	29
31	37	41	43	47	53	59	61	67	71
73	79	83	89	97	101	103	107	109	113
127	131	137	139	149	151	157	163	167	173
179	181	191	193	197	199	211	223	227	229
233	239	241	251	257	263	269	271	277	281
283	293	307	311	313	317	331	337	347	349
353	359	367	373	379	383	389	397	401	409
419	421	431	433	439	443	449	547	461	463
467	479	487	491	499	503	509	521	523	541
547	557	563	569	571	577	587	593	599	601
607	613	617	619	631	641	643	647	653	659
661	673	677	683	691	701	709	719	727	733
739	743	751	757	761	769	773	787	797	809
811	821	823	827	829	839	853	857	859	863
877	881	883	887	907	911	919	929	937	941
947	953	967	971	977	983	991	997	1009	1013

Squares and Square Roots

No.	Square	Square Root	No.	Square	Square Root	No.	Square	Square Root
1	1	1.000	51	2,601	7.141	101	10201	10.050
2	4	1.414	52	2,704	7.211	102	10,404	10.100
3	9	1.732	53	2,809	7.280	103	10,609	10.149
4	16	2.000	54	2,916	7.348	104	10,816	10.198
5	25	2.236	55	3,025	7.416	105	11,025	10.247
6	36	2.449	56	3,136	7.483	106	11,236	10.296
7	49	2.646	57	3,249	7.550	107	11,449	10.344
8	64	2.828	58	3,364	7.616	108	11,664	10.392
9	81	3.000	59	3,481	7.681	109	11,881	10.440
10	100	3.162	60	3,600	7.746	110	12,100	10.488
11	121	3.317	61	3,721	7.810	111	12,321	10.536
12	144	3.464	62	3,844	7.874	112	12,544	10.583
13	169	3.606	63	3,969	7.937	113	12,769	10.630
14	196	3.742	64	4,096	8.000	114	12,996	10.677
15	225	3.873	65	4,225	8.062	115	13,225	10.724
16	256	4.000	66	4,356	8.124	116	13,456	10.770
17	289	4.123	67	4,489	8.185	117	13,689	10.817
18	324	4.243	68	4,624	8.246	118	13,924	10.863
19	361	4.359	69	4,761	8.307	119	14,161	10.909
20	400	4.472	70	4,900	8.367	120	14,400	10.954
21	441	4.583	71	5,041	8.426	121	14,641	11.000
22	484	4.690	72	5,184	8.485	122	14,884	11.045
23	529	4.796	73	5,329	8.544	123	15,129	11.091
24	576	4.899	74	5,476	8.602	124	15,376	11.136
25	625	5.000	75	5,625	8.660	125	15,625	11.180
26	676	5.099	76	5,776	8.718	126	15,876	11.225
27	729	5.196	77	5,929	8.775	127	16,129	11.269
28	784	5.292	78	6,084	8.832	128	16,384	11.314
29	841	5.385	79	6,241	8.888	129	16,641	11.358
30	900	5.477	80	6,400	8.944	130	16,900	11.402
31	961	5.568	81	6,561	9.000	131	17,161	11.446
32	1,024	5.657	82	6,724	9.055	132	17,424	11.489
33	1,089	5.745	83	6,889	9.110	133	17,689	11.533
34	1,156	5.831	84	7,056	9.165	134	17,956	11.576
35	1,225	5.916	85	7,225	9.220	135	18,225	11.619
36	1,296	6.000	86	7,396	9.274	136	18,496	11.662
37	1,369	6.083	87	7,569	9.327	137	18,769	11.705
38	1,444	6.164	88	7,744	9.381	138	19,044	11.747
39	1,521	6.245	89	7,921	9.434	139	19,321	11.790
40	1,600	6.325	90	8,100	9.487	140	19,600	11.832
41	1,681	6.403	91	8,281	9.539	141	19,881	11.874
42	1,764	6.481	92	8,464	9.592	142	20,164	11.916
43	1,849	6.557	93	8,649	9.644	143	20,449	11.958
44	1,936	6.633	94	8,836	9.695	144	20,736	12.000
45	2,025	6.708	95	9,025	9.747	145	21,025	12.042
46	2,116	6.782	96	9,216	9.798	146	21,316	12.083
47	2,209	6.856	97	9,409	9.849	147	21,609	12.124
48	2,304	6.928	98	9,604	9.899	148	21,904	12.166
49	2,401	7.000	99	9,801	9.950	149	22,201	12.207
50	2,500	7.071	100	10,000	10.000	150	22,500	12.247

Fraction/Decimal Equivalents

Fraction	Decimal	Fraction	Decimal
$\frac{1}{2}$	0.5	$\frac{5}{10}$	0.5
$\frac{1}{3}$	0.3	$\frac{6}{10}$	0.6
$\frac{2}{3}$	0.6	$\frac{7}{10}$	0.7
$\frac{1}{4}$	0.25	$\frac{8}{10}$	0.8
$\frac{2}{4}$	0.5	$\frac{9}{10}$	0.9
$\frac{3}{4}$	0.75	$\frac{1}{16}$	0.0625
$\frac{1}{5}$	0.2	$\frac{2}{16}$	0.125
$\frac{2}{5}$	0.4	$\frac{3}{16}$	0.1875
$\frac{3}{5}$	0.6	$\frac{4}{16}$	0.25
$\frac{4}{5}$	0.8	$\frac{5}{16}$	0.3125
$\frac{1}{8}$	0.125	$\frac{6}{16}$	0.375
$\frac{2}{8}$	0.25	$\frac{7}{16}$	0.4375
$\frac{3}{8}$	0.375	$\frac{8}{16}$	0.5
$\frac{4}{8}$	0.5	$\frac{9}{16}$	0.5625
$\frac{5}{8}$	0.625	$\frac{10}{16}$	0.625
$\frac{6}{8}$	0.75	$\frac{11}{16}$	0.6875
$\frac{7}{8}$	0.875	$\frac{12}{16}$	0.75
$\frac{1}{10}$	0.1	$\frac{13}{16}$	0.8125
$\frac{2}{10}$	0.2	$\frac{14}{16}$	0.875
$\frac{3}{10}$	0.3	$\frac{15}{16}$	0.9375
$\frac{4}{10}$	0.4		